GW00992228

Alien Life Imagined

One day, astrobiologists could make the most fantastic discovery of all time: the detection of complex extraterrestrial life. As space agencies continue to search for life in our universe, fundamental questions are raised: are we awake to the revolutionary effects on human science, society and culture that alien contact will bring? And how is it possible to imagine the unknown?

In this book, Mark Brake tells the compelling story of how the portrayal of extraterrestrial life has developed over the last two and a half thousand years. Taking examples from the history of science, philosophy, film and fiction, he showcases how scholars, scientists, film-makers and writers have devoted their energies to imagining life beyond this Earth. From Newton to Kubrick, and from Lucian to H. G. Wells, this is a fascinating account for anyone interested in the extraterrestrial life debate, from general readers to amateur astronomers and undergraduate students studying astronomical sciences.

MARK BRAKE is a freelance academic and is well known for his work in communicating the relationship between space, science and culture. A former Professor of Science Communication, he founded the world's first undergraduate degree programme in astrobiology.

Alien Life Imagined

Communicating the Science and Culture of Astrobiology

MARK BRAKE

CAMBRIDGE UNIVERSITY PRESS
Cambridge, New York, Melbourne, Madrid, Cape Town,
Singapore, São Paulo, Delhi, Mexico City

Cambridge University Press
The Edinburgh Building, Cambridge CB2 8RU, UK

Published in the United States of America by Cambridge University Press, New York

www.cambridge.org
Information on this title: www.cambridge.org/9780521491297

First published 2013

Printed and bound in the United Kingdom by MPG Books Group

A catalogue record for this publication is available from the British Library

Library of Congress Cataloguing in Publication data

Brake, Mark.
Alien life imagined: Communicating the Science and Culture of Astrobiology / M Brake.
pages cm
Includes bibliographical references and index.
ISBN 978-0-521-49129-7 (Hardback)
1. Unidentified flying object literature. 2. Unidentified flying objects. 3. Voyages, Imaginary. 4. Life on other planets. I. Title.
TL788.7.B73 2013
576.8′39–dc23

2012022018

ISBN 978-0-521-49129-7 Hardback

Contents

1

Kosmos: aliens in ancient Greece

On our way we passed many countries and put in at the Morning Star, which was just being colonised. We landed there and procured water. Going aboard and making for the Zodiac, we passed the Sun to port, hugging the shore. We did not land, though many of my comrades wanted to; for the wind was unfavourable. But we saw that the country was green and fertile and well-watered, and full of untold good things.

Sailing the next night and day we reached Lamp-town toward evening, already being on our downward way. This city lies in the air midway between the Pleiades and the Hyades, though much lower than the Zodiac. On landing, we did not find any men at all, but a lot of lamps running about and loitering in the public square and at the harbour. Some of them were small and poor, so to speak: a few, being great and powerful, were very splendid and conspicuous. Each of them has his own house, or sconce, they have names like men, and we heard them talking. They offered us no harm, but invited us to be their guests. We were afraid, however, and none of us ventured to eat a mouthful or close an eye. They have a public building in the centre of the city, where their magistrate sits all night and calls each of them by name, and whoever does not answer is sentenced to death for deserting. They are executed by being put out. We were at court, saw what went on, and heard the lamps defend themselves and tell why they came late. There I recognised our own lamp: I spoke to him and enquired how things were at home, and he told me all about them.[1]

Lucian, *A True Story*, trans. A. M. Harmon, parallel English and Greek

The war of the worldviews

Astrobiology has some backstory.

The tale of our changing view of the possibility of life beyond the Earth is one that begins in ancient times, during the grandeur that was the ancient Greek world. In fact, the intellectual and cultural brilliance of the Greeks will have a profound effect on the rest of our narrative. But exactly what historical root does the plurality of worlds debate have in antiquity? What precursors in philosophy and literature, what flights of imaginative phantasy in fact and fiction? And what bearing did these ideas have on the science and culture of 'astrobiology' that was to develop over the coming millennia? Such are the questions that this first chapter seeks to address.

We shall discover the way in which philosophies of the classical world influenced succeeding ages, especially in relation to the natural sciences of astronomy and biology. The worldviews that shape our narrative are not dreamt up, out of thin air. So we shall be careful to trace the relevant movements in philosophy, culture, economy, and society for our story. For we need to know what prevailing conditions of culture and economy, of people and politics, led to such a remarkable history. Moreover, as it is the classical Greek world we have to thank for the conscious and unbroken thread of rational thinking, why *then*? Why in ancient Greece does such a reasoned philosophy develop, one so truly modern that its implications for life in the universe seem so contemporary?

The modern study of astrobiology makes use of many disciplines. As its main mission is to try explaining the origin, evolution, and distribution of life in the universe, its history often brings other fields into sharp focus. A history of astrobiology – naturally we have not always referred to the subject as such – must study advances in philosophy, physics, astronomy, cosmology, biology, and geology, among others, as rational speculation on extraterrestrial life has mostly concerned itself with hypotheses that fit firmly into existing scientific paradigms. So, the work of philosophers and scientists, especially astronomers, deserves special attention in this history, as their ideas about extraterrestrials have often had a strong influence on others.

The other primary goal of this book is to outline a history of fictional ideas about intelligent extraterrestrial life. As eminent writers are also intellectuals in the human family, their creative morphing of scientific ideas into symbols of the human condition

> is often an unconscious and therefore particularly valuable reflection of the assumptions and attitudes held by society. By virtue of its ability to project and dramatise, science fiction has been a particularly effective, and perhaps for many readers the only, means for generating concern and thought about the social, philosophical and moral consequences of scientific progress.[2]

Learning how poets react to extraterrestrial life, in other words, can teach us much about humanity.

Finally, there is the war of the worldviews. We shall discover that, throughout the ages, an evolving battle has played out between conflicting philosophies. At times of scientific revolution, this conflict peaks with great drama, such as the clash between Galileo and the Inquisition, and the controversy between Darwin and the creationists. But the schism is ancient in origin. Even in the ancient Greek world we find thought diverged into two paths, one materialist, one idealist.

And so to the ancient world, where we meet the first materialists in the Atomists, and the first idealists in Plato and Aristotle. With the Atomists we shall find a body of work that divines the shape of things to come: a worldview that embraced evolution, an atomic world of matter in motion, with no God. And with idealist thinkers Plato and Aristotle we encounter a philosophy of reaction, one developed out of the chronic fear of change when the strong slave-owning city-state of Athens was brought under the iron fist of Sparta.

The ancient Greek world

The central period of Greek thought hauls us into a history of the alien. With a history as glorious as the Greek, the puzzle is where to begin. But two crucial vectors make that choice a little easier. First is the dramatic rise of the city-state of Athens, and its associated empire. Second is the parallel development in philosophy. For in these two factors we can identify the beginning of a worldview that came to dominate the idea of life in the universe for two millennia.

The irresistible rise of Athens emerged after a Greek victory in the Persian Wars in 479 BC. The city burgeoned into the cultural and economic epicentre of the Greek world. With audacity and drive, the Athenians had fought off Persia's imperial aggression. Monetary and military savvy had proved priceless. Themistocles, the most prominent politician in Athens, had persuaded his fellow citizens to plough profits back into the push against the Persians. The coinage used to construct the city's commanding armada came from the Laurion silver mines, one of the chief sources of income for the Athenian state, and notorious for the treatment of the slaves who mined it. The poor powered the navy. This ensured victory for the city, and the support of the common people for its government.

Athens became a beacon. For the next hundred years, despite losing the war with Sparta, the city was the intellectual centre of the Greek world. Its fame was such that the literati flocked to the city. Artists and sculptors, philosophers and historians alike were drawn as scholars to the flame. Of the philosophers who came, we can identify two main tendencies of thought: those thinkers who believed in a worldview in which *matter* is primary, and those whose view was that *ideas* make up either all, or a major part of, the world of matter.

Table 1.1. *Two traditions: philosophy and science in the ancient Greek world*

		Philosophy and science	
Date	Social developments	Materialist tradition	Idealist tradition
600 BC	Age of tyrants	Influence of ancient learning	
		Thales, *Anaximander*, the Ionian materialist theory of nature	
	Persian conquest of Ionia	*Pythagoras*, maths and physical law	
		Heraclitus, philosophy of change	
	Liberation from Persians		
500 BC	Mining and metal-working	*Philolaus*, spherical Earth in orbit	
	Shipbuilding	*Empedocles*, four elements	*Parmenides*, change illusory
	Peloponnesian War	*Leucippus*, *Democritus*, atomic theory	
	Athenian democracy		
400 BC			*Socrates*, the dialectic method
	Defeat and reaction in Athens		*Plato*, Idealism
		Heraclides, nongeocentric cosmos	*Eudoxus*, heavenly spheres
	Triumph of Macedon		*Aristotle*, descriptive biology
	Alexander's conquests		
300 BC	Museum of Alexandria	*Epicurus*, atomic philosophy	
	Hellenistic influence spreads abroad		*Euclid*, geometry
	Wars with Carthage	*Aristarchus*, rotating Earth, heliocentrism	
200 BC	Great spread of slavery		*Hipparchus*, epicyclic cosmos
100 BC	Roman civil wars		
	Conquest of Gaul	*Lucretius*, atomic materialism	
0 BC	Jewish revolt	*Hero*, mechanics, steam engine	

Table 1.1. (cont.)

		Philosophy and science	
Date	Social developments	Materialist tradition	Idealist tradition
	Spread of Christianity		
AD 100	Marcus Aurelius, philosopher emperor		*Ptolemy*, descriptive astronomy
		Lucian, 'A True Story' fiction	
AD 200	Decline of city economy and trade		
	Crises and barbarian invasion		

Curiously, both tendencies owed a major part of their roots to the tradition of Ionian philosophy, particularly Pythagorean astronomy. And both worldviews were set to the task that was to become the litmus of wisdom for the Athenian age: to explain the motions of the Sun, Moon, and planets that whirled above their empire and the rest of the ancient world. The imagined solutions to this task held implications for the prospects of life beyond the Earth.

Of the two main philosophical tendencies, the more enlightened belonged to the Atomists. They envisaged a world ruled only by matter in motion, a cosmos that evolved in time. Inspired by early notions of evolution, they imagined nothing but atoms and void, and, more than two millennia before H. G. Wells' *The War of the Worlds*, took life into deepest infinite space, in fact and fiction.

The other tendency was that of the philosophers of reaction, those who opposed the idea of a plurality of worlds. Led by Plato and Aristotle, this tradition of the idealist doctrine was to become the dominant worldview. For their teachings formed the basis of the European philosophical tradition, with philosophy becoming 'a series of footnotes to Plato',[3] and science 'a series of footnotes to Aristotle'.[4] And, in the hands of the medieval Church, their ideas were effectively used to hold back modern science, including astronomy and biology, for two thousand years.

Back to the Brotherhood

Our journey through a history of the philosophy and culture of Greek science must include a backstory: the rise of the Pythagorean Brotherhood. The very word 'philosophy' is Pythagorean in provenance. When we use the word 'harmony' in its broader sense, when we call numbers 'figures', we speak the language of the Brotherhood. And their approach was the dawning of a new era; through their application of mathematics to the human experience of the natural world, they were among the founders of what we understand today as science.

The Pythagorean Brotherhood was founded in that remarkable century of awakening, the sixth century BC. It was not only in Greece that such philosophers were to be found. Elsewhere in the world, the eruption of the Iron Age gave scope to men with similar imaginings and messages. The inspired spiritual teachings of Siddhārtha Gautama in ancient India led to the foundation of Buddhism. In Palestine were the prophets and the later authors of the Wisdom literature, such as Ecclesiastes and the book of Job. And the teachings of Chinese thinkers, such as Confucius and Lao-tze, had begun to deeply influence Eastern life and thought.

It was an interconnected world. Indeed, in Egypt Jeremiah may well have met with Thales, regarded by Aristotle and by Bertrand Russell alike as the first philosopher in the Greek (and Western) tradition. Thales had hailed from Miletus, an ancient city on the western coast of what is now Turkey. It was a seat of considerable learning, one that produced notable ancient philosophers. Here also was born the atheist Anaximander, the teacher of Pythagoras, who was himself born sometime between 580 and 572 BC on the Greek island of Samos, set in the North Aegean Sea, close to Miletus.

There is some dispute as to whether Pythagoras truly was an entirely legendary figure. But that need not concern us here. The Brotherhood that bore his name was to have a great influence on later Greek cosmology and thought, chiefly through its most prominent exponents, Plato and Aristotle, but also through their opponents, the Atomists. Pythagorean astronomy was part of an all-embracing philosophy. The Brotherhood tried to synthesise a holistic view of the universe, one that incorporated religion with science, medicine with cosmology, mathematics with music; mind, body, and spirit as one.

Very little biographical detail is known about Pythagoras. None of his writings survive. So it is not possible to say whether some aspect of the Pythagorean worldview was the work of Pythagoras himself, or that of one of the Brotherhood. Much of the philosophical thought associated with Pythagoras may have been those of his brothers or successors. And this is quite apt when you consider that for many Pythagoreanism was the first expression of collective, democratic thought. Nor is it entirely accurate to describe the School as a brotherhood. Women were given opportunity to study; though they were also encouraged to learn domestic skills as well as philosophy, as they were considered different from men, in good ways as well as bad.

In the West at least, Pythagoras appears to have been among the first to call himself a philosopher, a lover of wisdom. Around 530 BC, he travelled south from Samos to Croton, in southern Italy. There he founded a secret school of religious philosophy. His reputation seems to have preceded him. For soon after his arrival, the Brotherhood ruled the town, and presently dominated a major part of *Magna Grecia*. From the eighth century BC, this area of *Greater Greece* was the region in southern Italy and Sicily that was colonised by Greek settlers, who brought an enduring stamp of their civilisation. Alas, in the case of the Pythagoreans, their secular sway was fleeting. They were banished from Croton, their temples razed to the ground, and members of the Brotherhood butchered.

Both *mathematical* and *mystical*,[5] Pythagorean philosophy was a blend of two tendencies. Many are understandably sceptical as to whether Pythagorean mathematics was justly original. The famous theorem on the right-angled triangle, for example, was known to have served the Egyptians and Babylonians way before the Brotherhood was born. And there is evidence to suggest that the entire number theory of the Pythagoreans, in both its mystical and mathematical aspects, was drawn from Eastern thought. But synthesis was paramount. The Brotherhood created a fusion of mathematics and philosophy that has had a lasting influence on the development of science.

The cosmology of the Pythagoreans

The Brotherhood saw that numbers were the key to comprehending the cosmos. In this, it is vital to understand that they were not limiting, or reducing, the human experience. For them, philosophy began in wonder. It was the highest music, and the highest form of philosophy concerned itself with numbers, for ultimately numbers were at the root of all things. When all was told, when logical thought had done its best, the wonder would remain. Rather than mathematics reducing the human experience, it enriches it.

The Pythagorean idea of 'harmony' is a case in point, an example of the way in which the Brotherhood orchestrated a unified view of their universe. Numbers were not thrown into the world by chance. They were ordered, or ordered themselves, like a musical scale, like the structure of a crystal, according to the universal laws of harmony. This basic Pythagorean idea of *armonia* saw the human frame as a kind of musical instrument. Each string within must have the right tension, the correct balance, for the human soul to be in tune. And these musical metaphors are still used in medicine. We talk of 'tone' and 'tonic', of the body being 'well-tempered' – all part of our Pythagorean heritage.

This Pythagorean focus on *armonia* was projected from the body and soul of man, out to the stars that seemed to circle the Greek world. The ancient

Babylonians and Egyptians had thought of the universe as a kind of cosmic oyster. It was an ancient cosmology with water underneath, and water overhead, but supported by a solid firmament. And what the Egyptians and Babylonians began, the Greeks refined. With some finesse, they developed a rational, rather than mythological, cosmology.

So it was that Anaximander of Miletus, Pythagoras' teacher, and one of the foremost Ionian philosophers, had visualised one of the first mechanical models of the universe. Anaximander's cosmos was not the closed oyster of the ancients. It was a universe infinite in space and time. The matter of this universe was not ordinary material, but indestructible, and eternal. And out of this stuff, all things were made, and into it they returned, as if in some cosmic recycling scheme.

Anaximander set the Earth adrift. True, like the cosmologies of other ancient cultures, Anaximander's system was geocentric, or Earth-centred. But rather than picturing an earth-disc floating in water, Anaximander rather curiously considered the Earth to be a cylindrical column, adrift in space. It was a bold innovation. For he surrounded this cylindrical Earth with air, floating upright at the centre of the universe, without support or structure. And yet the Earth was not in freefall. Since it was at the centre of all things, it had no favoured way in which to fall. For its falling would upset the symmetry and balance of such an ordered cosmos.

The Earth became a sphere. The Pythagoreans transformed this cylindrical Earth of Anaximander's into an orb, and around this central Earth, moving in concentric circles, revolved the Sun, Moon, and planets. Each heavenly body was fixed to a sphere, and the fleet revolution of this ingenious system inspired a musical resonance of the air. And so, claimed the Brotherhood, each planet would hum with its own unique pitch, for its music would depend on the ratio of its particular orbit about the Earth. Like a huge and heavenly musical instrument, the orbiting planets too were subject to the universal laws of harmony.

The Pythagorean harmony of the spheres lives on to this day. In his *Natural History* (*circa* AD 77), the Roman scientist and nobleman Pliny the Elder told that the Pythagoreans considered the musical interval formed by the Earth and Moon to be a tone; Moon to Mercury, a semi-tone; Mercury to Venus, a semi-tone; Venus to the Sun, a minor third; Sun to Mars, a tone; Mars to Jupiter, a semi-tone; Jupiter to Saturn, a semi-tone; and Saturn to the fixed stars, a minor third. The 'Pythagorean Scale' created from this musical arrangement is still recognised. And Pliny's report reveals not only a heavenly musical scale, but also a cosmic architecture that was to have a profound influence on the history of astrobiology. The story goes that only the master, Pythagoras, was graced with the gift of actually hearing this harmony of the spheres.

The music was an integral part of a cosmology that the Pythagoreans had evolved from ancient and avid sky-watchers who had scanned the stars and mapped the heavens. From such ancient observations, the Pythagoreans inherited a firmament. The stars were fixed and stationary, like pinholes in a dark fabric through which a cosmic fire could be seen. Perhaps they derived a sense of safety and security from the dependability of such a heavenly system. Immutable and predictable through the ages, the stars in their fixed and imagined constellations must have contrasted greatly with the turbulent lives of the sky-watchers below.

The firmament was not the only fascination the Pythagoreans inherited from the ancients. A number of so-called vagabond stars had in equal measure beguiled and perplexed sky-watchers for many millennia. To the ancient eye, without the use of a spyglass, only seven of these 'wanderers', or 'planets' as they were known, could be seen among the thousands of lights that bejewelled the firmament. The 'wanderers' were different. True, like the fixed stars, the Sun, Moon, Mercury, Venus, Mars, Jupiter, and Saturn all seemed to revolve once a day around the Earth. But the planets also had a peculiar motion.

The seven planetary bodies wandered and drifted along the path they traced across the night sky. Yet they did not roam about the entire heavens. Their cryptic conduct was restricted to a narrow strip of sky, a belt that encircled the spinning globe of the Earth at an angle of about twenty-three degrees to the equator. This belt, the Zodiac, was diced up by the ancients into a dozen divisions, and each division named for the constellation of fixed stars in that region of the Zodiac. Along this belt the planets roamed.

For those skywatchers more mythologically inclined, the passing of a planet through a constellation of the Zodiac had a double meaning. It provided data for their exacting observations, but it also supplied symbolic messages of ritual significance. For the more rational Pythagoreans, with their rule of harmony and number, a structure of the universe emerged.

In the Pythagorean system, the planets sped in circular orbits about the central Earth, and with the same reliability as that of the rotating sphere of fixed stars beyond. This scheme gave good account of the Sun's behaviour on its yearly journey through the plane of the ecliptic, the seeming path of the Sun across the sky. The Pythagorean system also gave a reasonable account of the rather less regular motion of the Moon. But the plain circular orbits came nowhere near explaining the observed motions of the other five wandering planets, of which more later.

Finer details notwithstanding, once the position and shape of the orbits were established, an ancient Solar System emerged. Jupiter and Saturn sketched a slow motion across the sky, seeming to keep up with the fixed stars beyond.

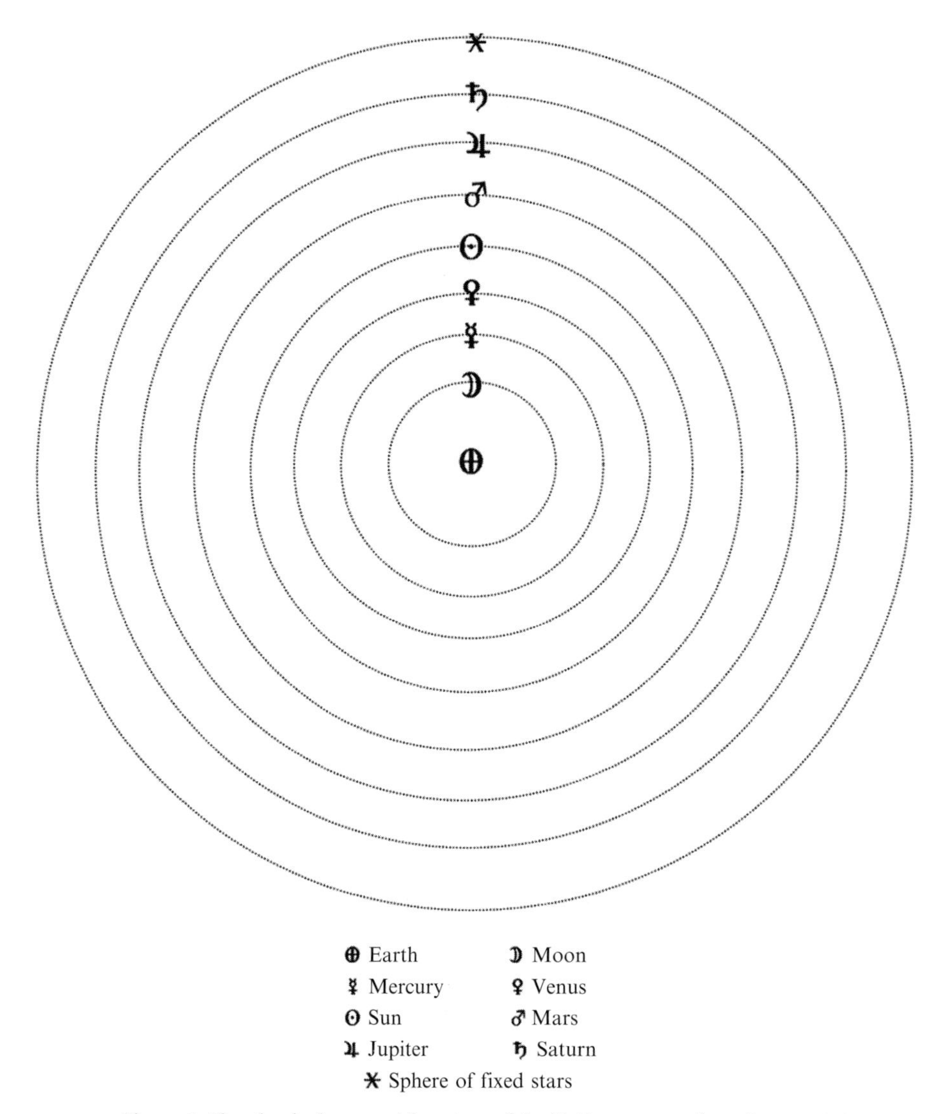

Figure 1 The classical geocentric sytem of the Pythagoreans; the spheres rotate counter-clockwise, as viewed from above.

Modern stargazers know that Jupiter takes roughly twelve years and Saturn thirty years to make a complete orbit about the Sun. To the ancient eye, since these planets kept pace with the stars, they were thought to be far from Earth, and close to the stellar sphere, which bounded their universe. At the other end of this heavenly scale was the Moon. Since our satellite loses twelve degrees each day in its apparent race with the stars, the Pythagoreans were justified in suggesting the Moon was close to Earth, which was thought stationary, and at the centre of the system.

Their cosmos was complete. Its outer limit was the stellar sphere. Just inside was Saturn, since it was the planet that took longest to move around the Zodiac. Next came Jupiter and Mars, set in order of decreasing orbital period, the time taken to make one complete orbit about the central Earth. Innermost was the Moon, since the lunar orbit placed it closest to us.

The remaining three planets of Sun, Venus, and Mercury, posed a problem. All three vagabond stars made their seeming journey about the Earth in the same common time of one year. But their order could not be unpicked in the same way as the other planets. Indeed, there was much debate on the matter among the philosophers. The layout shown in Figure 1 – Sun outside, Venus and Mercury within – was disputed, but for reasons now lost in the mists of time was the most popular order.

The cosmic system of the Pythagoreans had considerable potency and power.

Consider its layout. The system encapsulates most of the understanding enjoyed by the ancient astronomer. The term 'inferior planet' was now introduced for those bodies (Mercury and Venus) that came between the stationary Earth and the orbiting Sun. And the term 'superior planet' was used for those bodies (Mars, Jupiter, and Saturn) that lay beyond the Sun's orbit. The system gave no idea of the sheer size of the orbits, and no account of the inconsistencies of the planets in their apparent motion. But these more mathematical features were to develop later.

The layout also proved a potent tool for research. To begin with, the orbits embodied in the system gave good account of both the Moon's phases and eclipses. Indeed, the Pythagoreans had used lunar eclipses as the reason for believing that the Earth was spherical; the shadow cast by our world upon the Moon during these events was round, and, after all, the only solid that consistently projects a round shadow is the sphere. In due course, thinkers who followed in the wake of the Brotherhood used the ideas of the Pythagorean model to measure their universe with quite startling accuracy. Eratosthenes, for example, measured the Earth's circumference with unerring precision, during the third century BC. And during the second century BC, Aristarchus, of whom more later, ingeniously worked out the sizes and distances of the Sun and Moon.

Such stories give a primer of the power and ingenuity of ancient Greek astronomy, inspired by the Pythagoreans. Soon we shall consider their impact on thoughts of life elsewhere in the universe. But for now the potency of this simple cosmic model should be given more thought. The system made common sense for simple naked-eye observations that were made nightly, under a Greek sky that seemed to revolve about a central Earth. The model could predict important events. Eclipses came under its ken; those ominous cosmic encounters, which previously had given the ancient mind such pause for superstitious thought. It is hardly surprising, then, that the system became such a rational weapon, a conceptual tool with an increasing hold upon the minds of astronomers and laymen alike.

But what thoughts did the Pythagoreans themselves have on the question of life beyond the Earth? The early history of the Pythagoreans is somewhat conjectural, including their view on the plurality of inhabited worlds. It may be that Pythagoras held a Milesian view of plurality.[6] The Milesian school was founded around the Ionian town of Miletus in the sixth century BC, and included philosophers such as Thales, and Pythagoras' teacher Anaximander.

Anaximander himself was an early pluralist. According to Simplicius of Cilicia,[7] Anaximander held that there was indeed a multiplicity of worlds, similar to the Atomists Leucippus and Democritus, and later Epicurus. He supposed that worlds appeared and disappeared for a time, and that some were born while others perished. This movement was eternal, 'for without movement, there can be no generation, no destruction'.[8] The Roman philosopher Cicero wrote that Anaximander attributed different gods to the countless worlds of the heavens, 'For Anaximander, gods were born, but the time is long between their birth and their death; and the worlds are countless'.[9]

Another notable source on the pluralism of Anaximander is the third-century theologian Hippolytus of Rome. Hippolytus' most notable work is the compendious polemical tome, the *Refutation of All Heresies*. The book catalogues heretical beliefs, both pagan and Christian, and was recovered intact in 1842 on Mount Athos, situated on the eponymous peninsula in Macedonia.

The *Refutation* includes some fascinating remarks on Anaximander's cosmology. Hippolytus accounts for the idea that Anaximander, who was famous for the concept of *Apeiron*, or 'the boundless', believed that from the infinite comes the principle of beings, which in turn derive from the heavens and the worlds.[10] The belief places Anaximander before the Atomists and Epicureans, in holding that an infinity of worlds came and went. In our timeline of the Greek history of pluralism, we shall see how some thinkers conceptualised a single world, such as Plato and Aristotle, while others instead speculated on the existence of a series of worlds, continuous or noncontinuous.

The rational responders

It is instructive to take some time out of our history and reflect upon those important questions we mentioned at the top of the chapter, the kind of questions that rarely get asked in conventional histories of science. Since the Pythagoreans had been among the first to consider the mathematical nature of the cosmos, we are intrigued by the question, 'why then?' Why in ancient Greece does rational thinking dawn? And what were the conditions of culture and economy that led to such a dawning?

Crucial to such questions is the foundation of Greek cities. The characteristics of science, those that differentiate it from other aspects of human social activity, such as religion or art, are concerned mostly with the tradition and technique of how to do things. A work of science is a recipe: it tells you *how* to do certain things, *if* you want to do them. Science is not just a question of thought. It is the process of thought continually carried into action, constantly refined by practice. This makes science very potent: it is crucially linked to the means of providing for human needs. And it also explains why science should not be studied separately from technique.

A study of history reveals the mutual nature of science and technique. New techniques give rise to new science, and new developments in science give rise to new techniques. According to one tale, for example, Pythagoras realised the link between music and maths on hearing a blacksmith at work. In earshot of the sweet sound of the smith striking the anvil, Pythagoras understood that such harmony must bear some relation in mathematics. He spent some time with the blacksmith, examining utensils and exploring the simple ratios between tools and tones.

With the development of cities, the Greeks faced the challenge of having to provide for human needs on a massive scale. And with that came the necessity of innovation, in science and technique. Once this provision had been met came the challenge of administration. In these ancient cities, the responsibility for this considerable administration had previously lain with the priests. They had sole access to the means of measuring, calculating and recording. Indeed, the term hieroglyphics means 'priest's writing'. It is a reminder that science has, from ancient times, been associated with a governing elite. In the minds of the great mass of people, a limited access to science provoked a deep mistrust of science, and book learning in general. Such was the case five thousand years ago with the first cities.

The Greek cities were truly revolutionary. Between the twelfth and sixth centuries BC, the Greeks forged a unique and more democratic culture. Theirs was the most telling exploitation of the evolving Iron Age. It is true that they were lucky in some respects. They were rather isolated from the conservative influence of the older civilisations. They were protected from invasion by their initial poverty, and later girded by their growing sea power. But crucially, they had *nous*. They were able to learn from the traditions of the more ancient civilisations, and took from these foreign cultures many practical techniques, as well as a number of diverse cosmologies.

The change to an Iron Age economy had been crucial. The throttlehold of the landed aristocracy was losing its grip. Power was being grabbed by a group of local bosses, with the help of the merchant class. This was a time of violent growth, with colonies materialising throughout the Mediterranean. So it makes perfect sense that the Ionian settlements of the eastern Aegean were at the forefront of the new philosophy, as this was also the epicentre of an economic revolution.

The Mediterranean was a cauldron of change, which catalysed an inspired approach to traditional questions. Rather than relying on ritual and superstition to fuel solutions, the Greeks worked a new paradigm. The worldview that emerged was both simple and material, based on daily life and labour. Those who pondered upon such matters were known as 'sophists', or wise men. In time they would become 'philosophers', lovers of wisdom, retailing the ancient knowledge for new markets.

Little of the early sophists is known. But it appears that groups of sages, as with the Brotherhood, set up religious orders that doubled as philosophic schools. Those that flourished advised democratic chiefs or tyrants (at that time the word tyrant carried no ethical censure) giving advice on a range of policy and topics. To be sure, it bestowed kudos on a regime to have an eminent sophist in tow.

So, the budding Iron Age civilisation crystallised a new social type in the sophists.[11] The fact that knowledge of these early thinkers has survived, that Raphael painted a masterpiece in their honour,[12] and that legends about their lives have lingered, shows just how significant they must have been in the ancient world. And democratic or aristocratic patrons notwithstanding, these thinkers were almost all affluent gentlemen.

And yet, this compelling rise of the philosopher was a global phenomenon. The impact of the Iron Age was felt strongly in many parts of the developed world. In ancient China, Confucius and Lao-tze acted as political or technical advisors. In early India lived the rishis and buddhas, Siddhārtha Gautama, the Buddha, being the most renowned. And as we have seen, in olden Palestine the prophets and the subsequent authors of the Wisdom literature, such as Ecclesiastes and the book of Job, were alive and writing. Many of these thinkers and workers consorted with princes and tried in vain to reform their governments. But the important point is this: they all shared an interest in formulating a worldview of man and nature.

The great triumph of this new social type in the philosopher is that it filled the gap in ideas, the cultural vacuum caused by the shift in the economy from Bronze to Iron Age. The sophists supplied knowledge for a new class of economy, one in which there was a new brand of ruler: princes, tyrants, and merchants. But most of these new philosophers were nothing like the great engineers of past regimes, who helped build pyramids, temples, and waterways. The sophists had little to do with the daily running of the economy. Rather, they were removed from this material side of government and so, for the greater part with the exception of the Atomists, their worldview was idealist, unable to progress practical science.

For a while at least, the Greek situation was unique. The early sophists of the Ionian and Milesian schools, and the Pythagorean Brotherhood, lived at a time before the slave state and the rule of the rich had fully taken hold.[13] So, the early Greek philosophers contrasted starkly with their eastern counterparts. They were at once atheistic, materialistic, and rational, focused less on ethics, and more on nature.

The Greeks did not invent civilisation, nor even inherit it. They discovered it. Only *they* appropriated the vast bulk of ancient learning that remained, after centuries of war and abandon in the old empires of Babylonia and Egypt. The Greeks pored over the ancient wisdom. What was obscure, they ignored; what was superstitious, they disregarded. With such keen intelligence, they forged a new culture, at once simpler, more abstract, and more rational.[14]

Such was the ingenious distillation of their classical culture. Greek civilisation emerged as one of the keystones of today's world culture. Its seminal influence is in political democracy and natural science, especially mathematics and astronomy. The Greeks had been impressed by the noble science of the older ancient civilisations; it influenced them greatly. The study of mathematics, astronomy, and medicine remained the basis of education throughout the classical Greek period and beyond, into the medieval. And even today, the ignoble sciences, such as chemistry and biology, have to struggle for the same cultural respect.

These were the conditions of economy and culture that led to the dawn of rational thinking. Consider the archetypal city-state of Attica, in which Athens was located. The region was not rich in corn, and was reliant upon exports of olive oil, silver, and pottery, to feed Athens' immense population of three hundred thousand. But the compact city was able to marshal its resources to the max. In such an environment, there were swift, if not violent, changes in economy and politics. Tradition was at a discount. Those citizens with impetus and nous were rewarded with an enhanced standing in society, free of the fetters of clan or tribe. Institutions and divinities became less significant; men became more important than gods.

The *dialectic* was born. The fading of ancient sanctions meant that each case had to be argued out on its merits. And the history of Greek science, or philosophy as it was in those days, is a study in a series of back-and-forth arguments, named the dialectic. The other vital factor is this: the very skill of making such arguments was greatly encouraged by the intense political culture of everyday

Greek life. The significance of deals in trade and law, where each man represented himself and judges were chosen at random, necessarily led to debates of the highest possible level. In such ways, the Greek compact city-states fostered superior prospects for the typical citizen, compared with the capitals of great empires.

And so the rational philosophy of the Greeks was nurtured. Now we see the Pythagoreans in a plain and clearer light. The Brotherhood rode on the first wave of Greek science, centred on Ionia, the land that felt the influence of the older civilisations most keenly. We can also view the Pythagorean School as one of the first expressions of democratic thought. Their worldview, apt for an age of progress, was positive and optimistic. Their philosophy was materialist. They debated on the makeup of the world, and how it had come to be. And their rationalism was that of a rising mercantile class, pitted against the orthodoxy of empire and landed nobility.

The legacy of the Brotherhood

The original Pythagoreans were both a political Brotherhood, as in the vein above, and a very influential academy of 'science'. For their politics, they were persecuted and dispersed. Some features of their order were like those of a religious school. Their way was ascetic, keeping property in common, and leading a communal life. Like a primitive Christian community, they obeyed ritual and restraint. They devoted time and energy to meditation and the contemplation of conscience. But the Pythagoreans were also a force in Italian politics. By some accounts, after their first sermon to the Crotonians of southern Italy, six hundred joined their communal life, without even returning home to bid farewell to their families.[15]

In philosophy, the Brotherhood understood that reality could be reasoned into number, a feat that represents nothing less than the beginning of mathematics and physics, at least in the West. As one scholar said, 'Pythagoras is the founder of European culture in the Mediterranean sphere'.[16] True, their mathematics was mystical, a form of idealism that lives on in the blessed trinity, the four evangelists, the seven deadly sins, and the number of the beast.[17] Then again, the mystical element of the Pythagorean 'harmony of the spheres' seems to have had rather capricious consequences.

Pythagorean harmony was a vital part of the work of the sixteenth-century German astronomer Johannes Kepler. As will be seen in due course, Kepler's mercurial talents made major contributions to the study of the heavens, and early medieval ideas of alien life. And the 'harmony of the spheres' also inspired Galileo's father, a lute-player and maker, Vincenzo Galilei. Vincenzo discovered a new mathematical relationship between string tension and pitch. One can only speculate as to whether Vincenzo's work on mathematical ideas in music convinced Galileo himself to take an interest in physics.

The Pythagorean system of a 'harmony of the spheres' was, at root, mathematical. As with the philosophers that followed, the Brotherhood sought the secrets of life in the universe, and understood the significance of the circle and the sphere. They knew the Earth to be a sphere, and believed the planets move in circular orbits. In this way, we owe them the very foundations of astronomy and cosmology.

So, for us, the philosophy of the Pythagoreans stands at a crossroads. And from it flows the trail of two very different systems of thought. One trail follows the abstract and mystic aspects of the Brotherhood. In their physics, the Pythagoreans often ran too far ahead of their actual experience. They substituted number mysticism for knowledge based on experiment. True, the Brotherhood founded the powerful mathematical tool of proof by deduction. But in science, deduction can be used to prove conspicuous claptrap from allegedly obvious principles.

These abstract and mystic aspects laid the foundation for the idealism of Plato and Aristotle. Along this trail we shall see that there is sociology as well as logic to science. Wrong ideas can spread widely, at least for a time. On this path, Plato divorces thought from experiment, and lays a curse upon the idea of pluralism that is not lifted for two millennia.

But along our first trail the contrast is stark. On this first path, the Brotherhood's number theory was to be given a materialist twist. Pythagorean ideas were developed further by Heraclides and Aristarchus to give an ancient picture of the cosmos close to our modern notion of the Solar System. And the ancient creed of Atomism, with philosophers such as Democritus and Epicurus, marshalled a material world celebrated in measure and number.

So, first to the Atomists' almost Darwinian worldview, where we encounter no God, a cosmos replete with alien life, an evolving universe, and a reality in which nothing exists but atoms and the void. We shall find in Atomism the inspiration for a form of ancient science fiction, and that good ideas can lie dormant for centuries, before finally igniting the scientific imagination.

Life in the cosmos

Our narrative so far has necessarily focused on developments in astronomy and cosmology. And the motive for this is plain enough. The early Greeks had inherited a body of knowledge that held a pecking order of priorities. A study of ancient history shows a sequence in which the disciplines of science were brought within the scope of human experience. Roughly, the order runs: mathematics, astronomy, physics, chemistry, and biology.

The order of this development is not easily explained. And yet, it seems to have been somewhat dictated by the needs of ruling or rising classes at various times. The control of the calendar, for example, a priestly occupation, led to a necessary focus on astronomy. It was two thousand years, however, before the ambition of budding manufacturers of the eighteenth century gave rise to modern chemistry.

The less prosaic science of biology was derived much later, and from the direct study of its subject matter. Even so, ancient ideas on biology were part of an evolving philosophy of the natural world. So let us dwell on that developing worldview, and the way in which ideas of the origin of life in the universe blend in with the budding paradigms.

When we consider the philosophy of the first Greek sophists, the Ionians, we find a worldview conceived from water. The Ionians held that the world was born of water, and from this aqueous origin came the elements: earth, air, and all living things. It is biogenesis, like the creation myth of Sumer, the earliest known civilisation, first settled in the late sixth millennium BC. Unlike the version of the same creation myth that appeared in the book of Genesis, the Sumerian myth is materialist in philosophy. It is a rational one, thought out by inhabitants of a delta country where good soil had to be nurtured from marshland.

The early Greek version was also worldly, with no mention of a creator. As the Ionian sophist Protagoras[18] was soon to say, 'Man is the measure of all things',[19] and like Laplace just as notoriously said of God centuries later to Napoleon, 'I have no need of that hypothesis'.[20] The atheistic and materialist paradigm of the Ionians was founded on a fascination with nature. For these early philosophers, all matter was alive. They had little need of the metaphysical. It is a worldview that was later held by a progression of philosophers in the same tradition, the Atomists among them.

Fascinated by fossils, Anaximander had taken the Ionian worldview a step further. His account of biogenesis had summoned the elements of earth, mist, and fire, from which the world was made. But he also wondered at the origin of life itself. Once more, the solution was aquatic. Anaximander reasoned that creatures long ago sprang out of the sea. The first beasts that came to life were enclosed in a bristly bark. Once ripe, the bark would dry out, and break up. And as the planet's humidity dissipated, dry land surfaced and, in time, man was forced to adapt.

Anaximander's early version of evolution is astonishingly perceptive. His account recognises the long infancy of humans, which may make them defenceless in a hostile, primeval world. So, to survive, Anaximander has humans evolve for some time inside the mouths of big fish, cosseted in the face of the Earth's hostile climate. In time they emerge, losing their scales. Third-century Roman scholar Censorinus gives another taste of the account:[21]

> Anaximander of Miletus considered that from warmed up water and earth emerged either fish or entirely fishlike animals. Inside these animals, men took form and embryos were held prisoners until puberty; only then, after these animals burst open, could men and women come out, now able to feed themselves.

It is easy to see why some regard Anaximander as evolution's most ancient champion. Among other things, his thoughts on an aquatic descent of man resurfaced centuries later as the aquatic ape hypothesis. It is true that Anaximander lacked a theory of natural selection. And yet his ideas hail from a time when we started explaining the natural world, without recourse to myth or legend. For a few, this is called the 'Greek miracle'. But there is no need of miracle here, any more than miracles are needed to explain the genesis of life.

So begins the idea of life in the cosmos. Not just here on Earth, but also out in space. The essential ingredient of wonder had been added to the mix. And this process, marking the very beginning of scientific thought, is born out of the material conditions of the time. Another essential ingredient was the question of change. For only with change can we begin to understand that the cosmos is everywhere in flux. Enter fellow Ionian Heraclitus, known by many as the very philosopher of change, as he is said to have taken as his motto: *panta rhei*, everything flows. Heraclitus believed fire to be the primary element, as fire's dynamic nature would lead to profound change, transforming all in its path.

Heraclitus cuts a curious character in cultural history. He became known as the 'weeping philosopher', due to his melancholia, and as such makes a cameo appearance in Shakespeare's *The Merchant of Venice* (1598). Three decades later, Flemish painter Johannes Moreelse portrayed Heraclitus, dressed in dark clothing, wringing his hands over the world; no doubt due to the common conception that Heraclitus had a poor opinion of human affairs.

We have Diogenes Laërtius to thank for these impressions of Heraclitus. Diogenes was a biographer of the Greek philosophers, who lived in the first half of the third century AD. It seems that Heraclitus was marvellously bright from childhood, a hint of the greatness to come, that he was born to an aristocratic family, yet that he abdicated a kingship in favour of his brother. It seems Heraclitus saw philosophy as a loftier cause.

His fixation on fire is quite telling: 'all things are an exchange for fire, and fire for all things, even as wares for gold, and gold for wares'.[22] This notion is a powerful one. It seems to be an ancient revelation of the law of the conservation of energy. And it also shows the way in which the sophists based their philosophy on the material existence of daily life; as in this case, with the interchange of ideas between physics and economy.

With Heraclitus was born the dialectic: the philosophy of opposites. For him, some things, such as flame, tend to move up, while others, like stone, tend to move down. This, Heraclitus called the 'upward-downward path'. Such behaviour carries on concurrently, and results in a 'hidden harmony'.[23] The two opposites are essential to one another. And as everything flows, with objects new from one moment to the next, it follows that one can never touch the same object twice. Each object ebbs and flows into renewal, a continuous harmony between a building up and a tearing down. Heraclitus named these oppositional processes 'strife', and held that the apparently stable state is a harmony of it, a clear ancestor of the idea of equilibrium.

The evolution of terrestrial life

Toward the end of Heraclitus' life, another materialist philosopher was born who was to have a significant impact on the question of the evolution of life in the cosmos. Empedocles is another sophist influenced by the Brotherhood, and Atomist in philosophy. He is best known as creator of the cosmogenic theory of the four classical elements: earth, air, fire, and water. For Empedocles, these basic elements became the world of phenomena, full of contrasts and oppositions.

Through experiment he showed that invisible air was also a material substance, and proposed the order of the ancient elements as earth, water, air, and fire. Each element was above the other, tending if disturbed to return to its place in the natural order of things. Empedocles also held that opposite properties, such as love and hate, were material tendencies that mechanically mixed and divided in a continuous process. These ideas bear a similarity with the Yin and Yang dualism of ancient China. Though probably independent in origin, Chinese dualism also held that two principles, such as fire and water, male and female, forge to form other elements. In Chinese, these were metal, wood, earth, and through further fusion the 'ten thousand things' of the material world.

The cosmogony of Empedocles incorporated the concept of a cyclical universe. It is pertinent here as it deals with the origin and development of life. In this cosmogony, Empedocles forged a theory of everything. His worldview describes the separation of elements, the formation of ocean and earth, of Moon and Sun, and of atmosphere. Even further, he accounts for the biogenesis of plants and animals, and for the physiology of humans. His cosmogony seems a little curious, but makes for an intriguing read, and is a crude but quirky anticipation of the theory of natural selection.

So let us consider evolution, Empedocles style. On initial combination, the elements produced some strange results: heads without necks, arms with no shoulders. In time, as these simple scraps of anatomy coalesced, horned heads on human bodies emerged. In the same way, bodies of oxen with human heads evolved, and creatures of twin sex. Generally, however, these peculiar products of natural forces perished as quickly as they came. More rarely, bits and pieces of anatomy adapted to each other, and more complex creatures lived on. So was the organic universe formed, sprung from spontaneous aggregations. The creatures created in this way were fit for each other, as if this had been the purpose. Presently, various factors separated the creatures of double sex into male and female, and the world was replenished with organic life.

The cosmogony of Empedocles is ballpark Darwinism. It is founded on the combination of his beloved elements. Indeed, Empedocles' passion for the elements appears to have trailed him to the grave. Biographer Diogenes Laërtius tells the legend that Empedocles perished by tossing himself into Mount Etna in Sicily. The plan was that his observing followers would believe his body vanished, and Empedocles himself reincarnated. Alas, the deceit was dramatically revealed when the active volcano spewed back one of his bronze sandals.

So we have a clear picture developing of these early Ionian sophists. Theirs was a dynamic worldview, one whose constituent material elements were in constant transformation. In short, the mutable elements were responsible for matter in motion, and for the development of life. It was a *paradigm* of change, created at a *time* of great change. Fittingly, their philosophy of opposites was itself subject to its own contradiction, as it had to meet two incompatible roles. First, their philosophy stood for the actual material world and its phenomena. Without resorting to gods, their ideas had to account for the daunting ancient landscape of land and sea, sunshine and storm.[24] Indeed, the idea of the fury of the elements lives on.

Second, however, the elements were also expected to represent quality. An account of characteristics, such as warm and cold, hard and soft, wet and dry, also had to be reckoned within the philosophy's compass. The elements had not yet been identified with specific material substances, of course; that occurred much later with the chemical elements of the nineteenth century. But, these early Greek philosophers did propose that the seeds of every element may be present in all things, a clear precursor of the idea of states of matter, such as solids, liquids, and gases.

Early Greek thinkers knew that change was key to the cosmos. Today, we understand that change is the basis of evolution. And that the lapse of time changes all things: language, culture, the Earth, the bounds of the sea, and science itself. Throughout nature, we see systems in motion, from the electron clouds that swarm the heart of the atom, to the gently wheeling galaxies in an expanding spacetime. In astrobiology too, change is the life force. It is the engine that drives mutation, the catalyst that sparks each revolution. The only human institution that rejects change is the cemetery. And yet change always has its enemies.

So it was with the Greeks. For in stark contrast to the idea of a material world in flux, is the philosophy that was to come after the Ionians, Pythagoreans, and Atomists. For the bulk of thinkers that followed, those in the idealist tradition of Plato and Aristotle, the watchword was stasis. And while they appropriated many ideas from the Greek pioneers, these later philosophers focused far more on the *static natural order* of the elements. Theirs was a fixed and immutable universe.

As we shall see later in the chapter, this idea of a static order became, in the hands of Aristotle, forcefully sanctified, and was used to hinder any efforts at progressive change, social change in particular. Indeed, taking the idea one step further in his *Republic*, Plato has the aristocracy rule through the 'noble lie'. Effectively, Plato's pretence was that God had made four classes of men, each equated to an element: the guardians and the philosophers, both made of gold, the soldiers, made of silver, and the common people, made of baser metals. As a result, the idea of a person 'proving their mettle' is with us still. The picture was clear: the perfect social world was one where the lower classes were inferior to the upper. This reflection, with the natural world allegedly mirrored by the social world, merely led to a misunderstanding of both. It twisted a good, materialist theory into a rather rigid and abstract one. And so the development of astronomy and science was held back for centuries, by saddling natural philosophy with fanciful notions of a universal and sanctified order.

The Atomists and life in the void

Thankfully, ancient science also had the Atomists.

The great triumph of Ionian philosophy had been to weave a worldview, one that accounted for the creation of the cosmos, and the way in which it worked. Given its time of conception, it is remarkable that a material worldview, without recourse to gods and design, was divined at all. Its great flaw, however, was its purely descriptive quality. It needed number. And it was this injection of quantity and number that was the enduring contribution of the Pythagoreans.

Mathematics was the means to a better model. Rather than being tied to the Brotherhood's idea of a cosmos of ideal numbers, the Atomists saw a universe made of tiny countless un-cuttable (*a-tomos*) particles. They used the number theory of the Pythagoreans, but cast away their mysticism. The *atom* was born. These particles moved through empty space, a movement that described all observable change. Naturally, the theory was far from perfect. The atoms were believed to be immutable, unalterable. But, after all, the Atomists lived at a time before gadgets, before the contraptions of discovery. Nonetheless, they understood that particles might explain nature's rich variety.

Their next stunning innovation was the *void*: the empty space through which the atoms moved, without artifice. Injecting the notion of nothingness into science was a daring move. Earlier thinkers had thought the cosmos a plenum, a space entirely replete with stuff. With what they regarded as a modicum of common sense, most prominent philosophers loathed the idea of a vacuum. For example, in complete conflict with the Atomists, Aristotle was later to decree *horror vacui*; nature abhors a vacuum.

Aristotle had good reason to reject a vacuum. In a complete void, he presumed, infinite speed would be possible, as movement would meet no resistance. Aristotle thought this prospect of infinite speed untenable, and so concluded a vacuum invalid. Later, as we shall see in due course, many of the major triumphs of Renaissance science, such as Galileo's dynamics, were realised through the overthrow of Aristotle's ideas.

The Atomist worldview had a progressive, political aspect too. Theirs was a world devoid of the preordained, a self-contained world, and one in no need of divine dabbling. For though early Atomism was deterministic, later Atomists allowed for variety and free will in human affairs. Plato and Aristotle had sufficient influence to stop the Atomist creed winning general support. But the credo lived on, a lasting heresy from the very early days of rational thought.

In the West, one of the foremost Atomists was a Greek thinker of the fifth century BC, named Democritus. If Heraclitus was the 'weeping philosopher', then Democritus was the 'laughing philosopher'. In the first century AD, the Greek doxographer[25] Sotion contrasted the two thinkers in a quirky coupling of laughing and weeping philosophers: 'Among the wise, instead of anger, Heraclitus was overtaken by tears, Democritus by laughter'.[26] And Italian Renaissance architect Donato Bramante was to paint the fresco 'Democritus and Heraclitus', housed in the Casa Panigarola in Milan.[27] Plato is said to have disliked Democritus so much that he wished his books burned.[28]

Records on the idea of atoms date back to ancient India, as well as ancient Greece. In India from around the sixth century BC, forms of atomism developed in the belief systems of Jain,[29] [30] Ajivika and Carvaka, the last a sceptical, materialistic, and atheistic school of thought. Later the Nyaya and Vaisheshika schools also refined theories about the way atoms may combine to form more complex objects. It is unclear whether Indian philosophy influenced Greek,[31] or vice versa, or indeed whether the two creeds evolved independently, though there is good reason to believe that Democritus travelled as far as India on his many voyages.[32]

Democritus had been born in the Ionian colony of Thrace, though many called him a Milesian.[33] His father was said to have been so wealthy that he received Xerxes, the King of Persia, on his march through Democritus' hometown of Abdera. To slake his thirst for knowledge, Democritus used his inheritance to travel, wandering far and wide through Egypt, India, and Ethiopia.[34] Indeed, Democritus declared that among his equals none had made such journeys, seen more lands, and met more scholars than himself.[35] Once back in his native land, Democritus then journeyed through the Greek world, to better understand its culture.

So began his writings on natural philosophy, including many references to other Greek philosophers (his wealth having enabled him to buy their writings on his travels), and most notably his greatest influence was Leucippus, the founder of Atomism. Hippolytus of Rome, the third-century theologian we spoke of earlier, recorded in his *Refutation of All Heresies* that the Atomists firmly believed that life was to be found elsewhere in the universe, and held that the cosmos contained many worlds:

> Democritus holds the same view as Leucippus about the elements, full and void . . . he spoke as if the things that are were in constant motion in the void; and that there are innumerable worlds, which differ in size. In some worlds there is no sun and moon, in others they are larger than in our world, and in others more numerous. The intervals between the worlds are unequal; in some parts there are more worlds, in others fewer; some are increasing, some at their height, some decreasing; in some parts they are arising, in others failing. There are some worlds devoid of living creatures or plants or any moisture.[36]

At heart, Atomism is the belief that anything might ultimately be made up of an amalgamation of tiny distinct pieces that can be divided no further. The power of the idea is that it might equally be applied to the groupings of society or space. Indeed, under his mentor Leucippus, Democritus may have been the first thinker to realise that the fuzzily luminous celestial body of the Milky Way would resolve into the light of discrete but distant stars. Aristotle disagreed, of course.

The Atomist version of cosmic pluralism, a belief in numerous other worlds, which harbour living extraterrestrials, had begun with the Ionians, particularly Anaximander. For some Atomists, in a universe replete with life, worlds would come and go, emerging and vanishing, some being born as others perished, in an eternal movement. For many other Greeks, remote stars had been of little concern. And to many they were merely small lights in a vaulted sky that lay beyond the 'sphere' of Saturn.

For the Atomists, the cosmic host of life-bearing worlds they visualised was beyond reach, however. By the term 'world', they (as well as almost all other ancient authors) did not mean a different Solar System, like our own, and potentially visible from Earth. So other 'worlds' were not extrasolar systems, planets in orbit about distant stars. Rather, each was a self-contained cosmos, like our own, with an Earth at the centre, and with planets and stellar vault surrounding. Maybe these 'worlds' could be called other 'realms', a lot like our contemporary 'multiverse' idea, that speculative set of other universes, invisible and unreachable from our own universe. And so these other worlds of the Greeks might be contemporaneous with their ancient world, or may form a linear succession in time.

Atomist pluralism was to meet its most exquisite expression through Epicurus. Though both his parents hailed from Athens, Epicurus was born around 341 BC on Samos, the home island of Pythagoras. Epicurus returned to Athens, where he founded The Garden, a school named for the garden he owned, and which made do as a meeting place for his fellow philosophers. Only a few fragments remain of Epicurus' 300 written works, and much of what we know of Epicureanism derives from later followers and commentators, such as Dante, in whose *Inferno* (Canto X, Circle Six, *Where the heretics lie*), Epicurus and his followers are condemned for their materialistic model. The Epicureans focused on friendship as a vital element of happiness. Their school appears to have been a moderately ascetic community, quite cosmopolitan by Athenian standards, including women and slaves, and practising vegetarianism.[37]

Epicurus championed the idea of a plurality of inhabited worlds in a number of writings, including a letter he wrote to one of his disciples, Herodotus. The letter begins with an outline of Epicurus' atomic theory, but soon illuminates his views on other worlds:

> Furthermore, there are infinite worlds both like and unlike this world of ours. For the atoms being infinite in number . . . are borne far out into space. For those atoms, which are of such nature that a world could be created out of them or made by them, have not been used up either on one world or on a limited number of worlds, nor again on all worlds which are alike, or on those which are different from these. So that there nowhere exists an obstacle to the infinite number of worlds.[38]

The above passage shows it is important to realise that Atomism was more than just a cosmology, and that the notion of pluralism was drawn from the logic of their worldview. So it is with the Epicurean idea of an infinite number of worlds, derived from an infinite number of atoms. The 'world' of the Epicureans was 'kosmos', a system of order, not of chaos. The kosmos was their observable universe. And they conjured up a stunning scheme of an infinite horde of such worlds, beyond the senses, but not beyond reason. The Atomist reasoning behind this picture was quite cultivated: as there must be an infinite number of atoms, and that an infinite number of atoms could not have been exhausted by our finite world, so, as our world was born by chance collision of atoms in motion, then other worlds must be forged in the same way.

That extraterrestrials inhabit these infinite worlds of Epicurus is made clear in a later section from the same source:

> Furthermore, we must believe that in all worlds there are living creatures and plants and other things we see in this world; for indeed no one could prove that in a world of one kind there might or might not have been included the kinds of seeds from which living things and plants and all the rest of the things we see are composed, and that in a world of another kind they could not have been.[39]

The worldview of Epicurus was godless.[40] Naturally, as it follows on from a cosmos where atoms and void are the sole constituents, and all things that pass do so through chance collision of such atoms. And so the plurality of worlds according to Epicurus was one born of atomic theory, with little time for theology:

> We must grasp this point, that the principal disturbance in the minds of men arises because they think that these celestial bodies are blessed and immortal, and yet have wills and actions and motives which are inconsistent with these attributes; and because they are always expecting or imagining some everlasting misery.[41]

One of the leading contemporary disciples of Epicurus was the fourth-century philosopher Metrodorus of Chios. Metrodorus was a sceptic, as well as an Atomist and pluralist, though he also seems to have held the rather esoteric theory that the stars are formed daily by the moisture in the air under the heat of the Sun. Nevertheless, Simplicius has Metrodorus quoting the following pluralist attitude, once more an advanced cosmology for the ancient world:

> It would be strange if a single ear of corn grew in a large plain or were there only one world in the infinite. And that worlds are infinite in number follows from the causes (i.e. atoms) being infinite.[42]

The Atomist worldview would in time spread like wildfire throughout Europe. It reached the Roman Empire, where its major messenger was the Roman poet and philosopher Lucretius. His famous didactic poem *De Rerum Natura* (*On the Nature of the Universe*) was an early exercise in science communication, a popularisation of the doctrine of the Epicureans. According to Lucretius, other worlds must exist, 'since there is illimitable space in every direction, and since seeds innumerable in number and unfathomable in sum are flying about in many ways driven in everlasting movement'.[43]

The description of life in the universe presented by Lucretius, which includes the origin of life, the origin of species, and human prehistory, is the longest and most thorough extant account from the ancient world. It represents something of a zenith in the rationalist and materialist tradition, one that has its roots in the pre-Socratic philosophers we spoke of earlier, but whose theories are now sadly fragmented. As we have seen, in the ancient world the main protagonists were, on one side, the Atomist, mechanist approach, such as that of Democritus and Epicurus, and on the other side Aristotle and Plato's teleology, the philosophy that holds that final causes exist in nature, and that the design and purpose found in human actions are also present in the rest of nature.

Of those earlier sophists, the zoogenic theory of Empedocles bears the closest resemblance to that of Lucretius. Both provide a mechanistic account of biogenesis, one that dispenses with the need for any design or divinity. Like Empedocles, Lucretius presented a theory that has been seen as a forerunner of the Darwin–Wallace theory of evolution. To be sure, Lucretius was the main target for attacks by creationists, until the publication of *On the Origin of Species* in 1859. He still has a place in the infamous and creationist 'black museums' of evolutionary thinkers.

It is instructive to let Lucretius speak for himself. The materialist approach is unmistakable in this wonderful quote from his *De Rerum Natura*:

> Take first the pure and undimmed lustre of the sky and all that it enshrines ... Desist from thrusting out reasoning from your mind because of its disconcerting novelty. Weigh it, rather, with discerning judgement. Then, if it seems to you true, give in. If it is false, gird yourself to oppose it. For the mind wants to discover by reasoning what exists in the infinity of space – that lies out there, beyond the ramparts of this world – that region into which the intellect longs to peer and into which the free projection of the mind does actually extend its flight ... Bear this well in mind, and you will immediately perceive that *nature is free and uncontrolled by proud masters* and runs the universe by herself without the aid of gods.[44]

Lucretius presents a staggering, Atomist vision of infinite space, colonised by countless stars and planets. His pluralism is striking in its godlessness, and a provenance due solely to the mechanism of its cosmology. Indeed, such conclusions would have been impossible, were it not for the physical principles of a materialist approach. Lucretius added that nothing of these worlds was unique, not even the Earth itself:

> It is in the highest degree unlikely that this earth and sky is the only one to have been created and that all those particles of matter outside are accomplishing nothing. This follows from the fact that our world has been made by nature through the spontaneous and casual collision and the multifarious, accidental, random and purposeless congregation and coalescence of atoms whose suddenly formed combinations could serve on each occasion as the starting point of substantial fabrics – earth and sea and sky and the races of living creatures. On every ground, therefore, you must admit that there exists elsewhere other congeries of matter similar to this one which the ether clasps in ardent embrace.[45]

Like Empedocles before him, Lucretius imagined a cosmogony, an ancient theory of everything. His worldview accounts for the nature of the universe, from cosmology to biogenesis: 'Turn your mind first to the animals. You will find the rule apply to the brutes that prowl the mountains, to the children of men, the voiceless scaly fish and all the forms of flying things. So you must admit that sky, earth, sun, moon, sea and the rest are not solitary, but rather numberless.'[46]

Set the controls for the heart of the Sun

As the likes of Lucretius made stunning Atomist pronouncements on life in the universe, a line of Pythagorean thinkers had made revolutionary steps in realising man's true place in the cosmos. The first of these, Philolaus, a contemporary of Democritus, and junior to Empedocles, was the earliest known thinker to assign motion to our planet. He argued that numbers governed the universe, but most significantly for our story, Philolaus is the originator in the West of the theory that the Earth was not the centre of the universe. Indeed, his notions of the cosmos are drastically different from what went before.

No longer did the Earth stand central, massive and immobile. Philolaus broke away from the geocentric tradition. At the hub of his 'world' was 'the hearth of the universe',[47] the Central Fire. This Central Fire could not be seen, as the civilised central region of the globe, the Greek world, was always facing away from the Fire, much as the dark side of the Moon was always facing away from the Earth. But beyond this cosy cosmos of the Central Fire was its main source of light, the 'outer fire', which bounded the universe on all sides. A wall of fiery ether, it was an eternal spring of luminous energy. The Sun served merely as its portal or lens, through which the outer light was drawn and dispersed. It is a fantastic notion. But perhaps no more incredible than what we believe today: the idea of a ball of burning gas, careering across an endless sky.

Philolaus set the Earth free. According to his system, the Moon was the sole object in the heavens considered similar to the Earth. Since it was bathed in sunlight for fifteen days at a time, Philolaus thought the Moon inhabited by life, fifteen times as strong as terrestrial creatures. The cosmology of Philolaus sparked a healthy debate in the nervous system of fellow philosophers. It would not be long before scholars realised the daily revolution of the entire sky was merely an illusion, one caused by the Earth's own motion.

Enter Heraclides, the next notable thinker in the Pythagorean tradition. Though, of course, he was not the only worker proposing such ideas at the time, Heraclides is credited by the doxographer Diogenes Laërtius as an important innovator of the cosmic model. For Heraclides understood that the apparent daily motion of the sky was created by the daily rotation of the Earth on its axis. The curious cosmology of Heraclides was a quirky Earth-centred cosmos, with a heliocentric twist. In a bold attempt to throw pure geocentrism into the dustbin of history, Heraclides had Mercury and Venus orbit the Sun, while the remainder of the system, Sun included, were satellites of the Earth.

This cosmic cunning is evocative of the life of its creator. The son of a wealthy nobleman, Heraclides was packed off to study at the Platonic Academy, under the founder Plato himself. Though he studied with Aristotle, when Plato briefly left for Sicily in 360 BC, he left his pupils in the charge of Heraclides. Praise indeed, especially when you consider Heraclides was allegedly one of the few philosophers who dared disagree with Plato.

The ideas of Philolaus and Heraclides were to dramatically influence Aristarchus of Samos, last in line of the Pythagorean astronomers. Hailing from Pythagoras' home island of Samos, this 'Greek Copernicus' dared to put the Sun and not the Earth at the centre of the universe. Although the original texts in which Aristarchus set out his system are lost, his references are impressive. The writings of Plutarch and Archimedes testify that Aristarchus proposed a heliocentric system. And Nicholas Copernicus was to resurrect, eighteen centuries later, this pinnacle of Pythagorean astronomy.

A younger contemporary of Aristarchus, Archimedes was one of the leading thinkers of antiquity: a great mathematician and engineer. One of Archimedes' most remarkable works is a short treatise called *The Sand Reckoner*, in which he tries to determine an upper limit for the number of grains of sand that the universe can hold. A rather quirky idea for a book. It is around only eight pages long in translation, and is dedicated to King Gelon II of Syracuse, a historic ancient city in southern Italy. Crucially, to meet the challenge he set himself, Archimedes used the system of Aristarchus:

> You King Gelon are aware the 'universe' is the name given by most astronomers to the sphere the centre of which is the centre of the Earth, while its radius is equal to the straight line between the centre of the Sun and the centre of the Earth. This is the common account as you have heard from astronomers.
>
> But Aristarchus has brought out a book consisting of certain hypotheses, wherein it appears, as a consequence of the assumptions made, that the universe is many times greater than the 'universe' just mentioned. His hypotheses are that the fixed stars and the Sun remain unmoved, that the Earth revolves about the Sun on the circumference of a circle, the Sun lying in the middle of the orbit, and that the sphere of fixed stars, situated about the same centre as the Sun, is so great that the circle in which he supposes the Earth to revolve bears such a proportion to the distance of the fixed stars as the centre of the sphere bears to its surface.[48]

The Greek historian and essayist Plutarch made the other outstanding citation of the system of Aristarchus. In his *On the Face in the Moon Disc*, Plutarch presents a long and curious treatise on topics such as the cause of the Moon's light, its peculiar colour, and the possibility of its being inhabited. But the dialogue also refers to Aristarchus' belief 'that the heavens is at rest, but that the Earth revolves in an oblique orbit, while it also rotates about its own axis'.[49]

Aristarchus was a materialist. His sole surviving work is a short treatise, *On the Sizes and Distances of the Sun and Moon*. And it is stunning. We spoke earlier in the chapter of science as more than a mere matter of thought. Rather, it is the process of thought carried into action. So it was with Aristarchus. He was not content to merely contemplate the cosmos; he wanted to get a real measure of it. And his treatise details ways to calculate not just the sizes of the Sun and Moon, but also their distances from the Earth, in Earth radii.

Using trigonometry, for example, Aristarchus found that the distance to the Sun was about twenty times more than the distance to the Moon. And since the apparent sizes of the Sun and the Moon in the sky are the same, the Sun must be twenty times bigger. The method is sound, it is the measurements that proved difficult. The impression we get of Aristarchus is of a rational worker, inventive in ideas and precise in practice. Astronomers used his elegant method of finding the distance of the Sun for centuries to come. If the results were wrong, it was only because he was working two thousand years before the invention of the spyglass.

So the radiant trail of cosmic thought that began with Pythagoras reached a luminescent peak with Aristarchus. Eighteen centuries before the Copernican dawn of the scientific revolution, Aristarchus imagined a Sun-centred universe.

Anaximander had set the Earth adrift, Philolaus had warmed the skies with his Central Fire, Heraclides had conjured up a cosmos with a heliocentric twist, and Aristarchus had delivered a Sun-centred system; all markers on the road to modern science. For scientists to embrace the search for habitable planets outside our Solar System, we first had to realise that such systems were solar, rather than terrestrial. The Sun had to become a star.

And yet Aristarchus' Sun-centred system did not succeed. It was not the materialist method that won out in the bid to become the dominant cosmology of the ancient Greeks. The Roman engineer and writer of the first century BC Vitruvius (he of Leonardo da Vinci's *Vitruvian Man*), said of Aristarchus that he was one of the leading astronomers of his time, a universal genius. We know better. For us, Aristarchus is the last in line of a flow of philosophers in the Pythagorean tradition.

Whither the Greek alien? Enter Plato's dark side

In spite of all this progress, science now began to fall into disgrace and decay. The materialist view of the cosmos that infused the philosophy of the Pythagoreans and the Atomists was a practical one. It was a philosophy of matter in motion, an account of nature and society from below and not above.[50] The most potent example of its power we have encountered was *On the Nature of Things*, Lucretius' Atomist poem. Centuries later, it still broadcasts the boundless energy of a world in flux, and man's power to alter it, by learning its laws.

But the greatest of Greek tragedies had already begun. Along the materialist path, there was only one step from Aristarchus to Copernicus, one step from Archimedes to Galileo, and only one step from the Atomists to Darwinian evolution. And yet it was Plato and Aristotle that came to the fore, with their robust defence of an idealist system that was already in ruin.

In later chapters, when we look at progress during the scientific revolution, we shall see how the Atomist and experimental science of Galileo fared against its main adversary, the idealism of Aristotle, espoused by the Church. And in the nineteenth century, this continuing conflict will be found in the enmity between biblical literalists on the one hand, and the scientists of the 'Darwinian revolution' on the other. At a deeper level, however, the progressing oppositions of this conflict hint at something other than just the science at stake. At heart, it is an indication of economic and political struggles. At each stage, idealism is used to profess that social problems are illusive, and materialism invokes practical tests of the real world and insists upon the need for change.

So it was with the Greeks. The idealism of Plato was in essence a reaction to the materialism of the Atomists. Plato's 'science' was steeped in politics. In both his philosophy and his physics, change was most foul. Truth, the Ideal, and Beauty were eternal, and beyond question. And as they were not to be found here on Earth, they were realised only in the immutable heavens. For Plato, the beauty of the stars was merely an aspect of the visible world. In 'truth', the stars were a diffuse copy of the real world of ideas. Any effort to understand them, or to calculate their motions, was thus absurd. In the words of Plato, 'Let us concentrate on (abstract) problems, said I, in astronomy as in geometry, and dismiss the heavenly bodies, if we intend truly to apprehend astronomy.'[51]

A century of civil war, and chronic civil unrest, had led to political and moral bankruptcy; economic hardship had made homeless many of its citizens. The classical Greek world confronted a stark crisis. In many ways, scholars regard it as one of the greatest failures in history:

> Plato and Aristotle each in his different way tries (by suggesting forms of constitution other than those under which the race had fallen into political decadence) to rescue the Greek world, which was so much to him from the political and social disaster to which it is rushing. But the Greek world was past saving.[52]

The crisis bled into Greek physics. Plato invented astrology. Ironically, the word means reasoning (*logos*) about the stars, whereas the old 'astronomy' had merely meant to order (*nomos*) them. Plato was seduced by the mystical beliefs of the Pythagoreans, particularly the cosmic significance of mathematics. But it was a knowingly abstract approach. And it took mathematics away from its practical origin and evolution, along a meagre and contemplative departure.

The common view of the heavenly bodies in the ancient Greek world, particularly the Sun, Moon, and planets, was that they were divine beings. A form of extraterrestrial life, perhaps, but not a rational one. This faith explains why old-fashioned folk were stunned by the impious Ionians, who had proposed the planets were just fiery globes, wandering an empty sky.

Plato went one step further. To the detriment of astronomy and the rational belief in life beyond the Earth, he fused mathematics and theology, and declared the planets divine. Their divinity was witnessed in their fixed and regular paths, orbits of perfect and circular motion. And just as a finger gently rubbed along the rim makes a wine glass sing, so the divine planets sang in their circular paths about the central Earth. Change was banished from the heavens, as Plato no doubt would have cast it out of society. The philosopher's highest calling, for Plato, became the consideration of perpetuity, his supreme pursuit the proof of man's immortality. For science, this meant oblivion. The developments in materialist philosophy now crumbled. The world of matter in motion, proposed by Pythagoreans and Atomists alike, their universe replete and seeded throughout with life, ground to a celestial and sterile halt.

It was based on this same pursuit of heavenly perfection that Plato concluded the shape of the world to be a perfect sphere. Similarly, all heavenly motion must be in a perfect circle, at uniform speed. It was all illusion: change without change, for motion in a circle returns to its origin. The task Plato now set for mathematicians and scholars was to dream up a cosmic model in which the apparent irregularities of the planets' paths across the Greek sky could be explained away by a system of regular motions in perfect circles.

Thus is the strange contradiction that was Platonic astronomy. His contribution to the field was trivial, and yet his influence was huge. His interest in the subject was slight, yet he held back science for two thousand years. Any valid physics of the stars was now held in stasis until the days of Copernicus and Galileo. That Plato's authority should have been so lasting is explained by a number of influences, but one factor stands apart: Aristotle.

Aristotle's lifeless universe

The ancient universe before the Greeks had been a kind of cosmic oyster, a cosmos all clammed up in space and time. The Pythagoreans had jemmied it open, and let a spherical Earth adrift, as the Atomists extended its outer limits to infinity, and seeded it with life. Theirs was a cosmos of change, a 'world' in motion. Aristotle now wound back the cosmic film, and returned an immobile Earth to its centre.

Witness, then, the Aristotelian universe: the closed cosmos resurrected, with hard limits in time and space. For two millennia his cosmology held sway. Aristotle's vision was of a two-tier, geocentric cosmos. The Earth, mutable and corruptible, was placed at the centre of a nested system of crystalline celestial spheres, from the sub-lunary to the sphere of the fixed stars. The sub-lunary sphere, essentially from the Moon to the Earth, was alone in being subject to the horrors of change, death and decay.

There was no room for the alien. For outside the Moon, the celestial or supralunary sphere, the cosmos was immutable and perfect. Critically, the Earth was not just a physical centre. It was also the centre of motion, and everything in the cosmos moved with respect to this single centre. Aristotle declared that if there were more than one world, more than just a single centre, elements such as earth and fire would have more than one natural place toward which to move, in his view a rational and natural contradiction. Aristotle concluded that the Earth was unique.

This is the system that was to stand the test of time. So let us consider this cosmology of Aristotle a little further, given its enduring importance that would

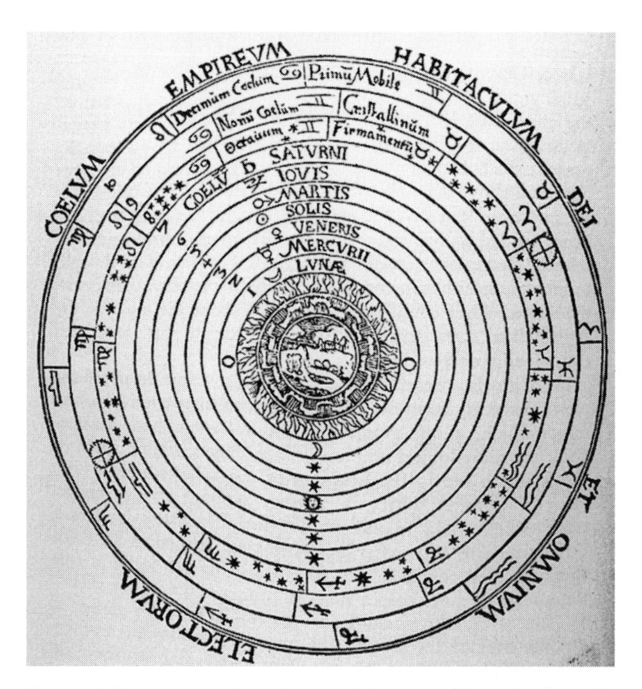

Figure 2 The geocentric universe of Aristotle. Note the four elements, nine celestial spheres, and the final sphere, Primum Mobile, driven by divinity. Beyond lay 'the habitation of God and the Elect'. From Peter Apian's *Cosmographia*, Antwerp, 1539.

last until the scientific revolution, and have a huge influence on medieval religion and culture. Aristotle's model is a masterly synthesis of earlier Earth-centred cosmologies. His walled-in universe encloses nine concentric spheres, clear as crystal. The outmost sphere is that of God, the Prime Mover. It is He who spins the cosmos from beyond. The movement He imparts to the outmost sphere is conveyed to each neighboring sphere in turn. Like a child's clockwork toy, the universe is reduced to a mechanical curiosity, with God keeping the machinery in motion.

In the Pythagorean version, Philolaus had filled the entire cosmos basking in the warmth of a Central Fire, a source of cosmic energy. With God's relegation to the outer limits, Aristotle now confers upon the Earth the most lowly and humble place in the whole universe: the centre. This inmost layer alone, the sub-lunary, is witness to dreadful change. Outside the sphere of the Moon, all is serene. Out in this supra-lunary region dwells each of the planets, orbiting with their God-given motion about the central Earth. Aristotle's refinement is the classic geocentric system of the ancients, dancing to Plato's divine tune of regular motion in a perfect circle. To the Greeks, this idea of the universe gifted cosmic comfort in frightening times. And later, to the dark medieval mind, the splitting of the cosmos into two, one part lowly, the other divine, one part change, the other immutable, brought an illusion of strength in times of turmoil.

The other genius of Aristotle is this: caught in crossfire of cosmologies, his model is that of supreme compromise. His vision is a fusion of worldviews, a synthesis of the material and the ideal, and a great consolation to the minds of the meek. His sub-lunary sphere is the realm of the materialists. It is the tier of Aristotle's cosmos in which the dynamic forces of Heraclitus are in constant flux. Here, matter is made up of different combinations of the four elements of Empedocles: earth, water, air, and fire. Each has a natural place to be, earth downward, fire upward, air and water horizontally. The elements are agents of change. They continually transmute, and the sub-lunary sphere is replete with their fusions. The composition of the atmosphere is not pure air, but a substance, a catalyst of change, which when set in train ignites to create meteors and comets. In short, the sub-lunary sphere is vibrant with life.

The rest of Aristotle's cosmos is dormant, divine but essentially dull. This is the realm of idealism. Here we find change to be an illusion, as with Plato's planetary orbits, as Aristotle conjures static perfection, writ large in the outer limits of his cosmos. For beyond the Moon, there is no change. The four earthly elements that catalyse change in the sub-lunary are gone in the supra-lunary sphere. As with Plato, the planets move about the central Earth, but it is movement without change. For circular motion is perpetual, without beginning or end, one continuous movement that returns into itself. The fabric of Aristotle's heavenly tier was quintessence, the fifth element. Chaste and immutable, it is flawlessly manifest in the shape of the crystal concentric orbs of planetary paths, as the sphere is the only perfect form. And the further out we fly beyond the Moon, the purer the quintessence becomes, until it meets its purest form in the sphere of Aristotle's God, the Prime Mover.

Aristotle considered the question of the plurality of other worlds most fully in two of his works, *On the Heavens* and *Metaphysics*. Within *On the Heavens*, he sets out his ideas about the four terrestrial elements, and a fifth celestial one, and the difference in their respective motions. Aristotle then muses on whether the cosmos can be infinite, concluding that it cannot, and going on to comment, 'This will lead us to a further question. Even if the total mass is not infinite, it may yet be great enough to admit a plurality of universes.'[53] This question then becomes the focus of two chapters, in the first of which Aristotle concludes there cannot be more worlds than one, as all movements of elemental particles of other worlds would be to our central Earth.

In the second relevant chapter of *On the Heavens*, Aristotle considers whether the cosmos, being a physical entity, should not have a multiplicity of embodiments. In other words, 'there either are, or may be, more heavens [or universes] than one'.[54] Aristotle tries to reason through this difficulty by suggesting, 'this world contains the entirety of matter'.[55] He elaborates in the following way:

> Now the universe is certainly a particular and a material thing: if however it is composed not of a part but of the whole of matter, then though the being of 'universe' and of 'this universe' are still distinct, yet there is no other universe, and no possibility of others being made, because all the matter is already included in this. It remains, then, only to prove that it is composed of all natural perceptible body.[56]

In order to prove his last point, Aristotle imagines a body beyond our universe. This body would have to be made of the four earthly elements, or of the special celestial matter that he says exists in the heavens with a circular motion. But he rejects such a proposition, suggesting no such matter can exist. It would be compelled to have the motions appropriate to each of those forms of matter, which are by their nature defined with respect to the position of the Earth, which is at the centre of the cosmos.

In his *Metaphysics*, Aristotle delivers another case against the idea of other worlds beyond ours. This argument is based on his contention that celestial movements have their origin in the Prime Mover, the arbiter of all motion in the cosmos. Aristotle's point is this: if there were more than one universe, by his own logic, there would have to be more than one Prime Mover. This idea Aristotle takes to be impossible as, 'So, the unmovable first mover is one both in definition and in number: so too, therefore, is that which is moved always and continuously; therefore there is one heaven alone.'[57] It is plain that Aristotle's stance on the question of other worlds was linked not just to his physics and cosmology, but also to his theology.

The *scala naturæ*

The description of Aristotle's two-tier universe given above is a strategic one. It is an intentional depiction of one tier in which 'everything flows', and another tier where 'nothing ever changes'. It signifies a period of great unrest in the Greek world, a time of heated debate. The lives of Aristotle, Epicurus the Atomist, and Zeno the founder of the Stoics, all crossed. So, our account of Aristotle's model gives the distilled essence of that debate, and the cunning way in which Aristotle steered a careful compromise through divergent ideas.

This Aristotelian universe was to prove vital to the drama of the Copernican and the Darwinian revolutions, and to the debate on the possibility of life beyond the Earth. Its toxic potential became more potent still when its two-tier simplicity grew into the multi-layered cosmos of the Latin medieval church. Crucial to our story later, the new system that emerged also embraced a form of life in the universe, both earthly and divine, in the shape of the *scala naturæ*. This Great Chain of Being, a grand scheme of the universe, had as its main feature a strict hierarchy, in which every object and every creature had its precise and allotted place.

Aristotle's compromise position on the question of change was far less damning, and more ambivalent, than Plato's. Aristotle shunned evolution and progress per se, and yet regarded change in nature as purposeful. As a keen naturalist, he saw movement in nature as a goal-driven process. The same was true for inanimate bodies. Just as a salmon swims upstream, so a stone will fall to Earth. Just as a spider weaves its web, so the flame of a fire will reach skywards. Each had their 'natural place' in the *scala naturæ*.

Plato's utter abomination of change resounded with the Church's belief in an unchanging heaven. He had made lowly the Earth, bound in a walled-in universe of nested spheres, made safe from uncivilised change. Aristotle's refinement was to have a ruinous result on the course of Western thought. For the *scala naturæ* also gifted the Church an ornate system of control; one does not desert one's place in the chain – that would be unthinkable.

As we shall see in the next chapter, the *scala naturæ* also became a system of life in the universe, both terrestrial and heavenly. It was to develop into an elaborate pageant of divine creation, an infinite procession of links, stretching from God down to the lowliest form of life. And so, the philosophy of Plato and the physics of Aristotle were to have a massive stranglehold on the development of ideas about life in the universe. Taken together, their philosophy provided a system that seemed to present a complete picture of the world, and a welcome solution in troubled times. Even as the cultural climate in Europe slowly evolved, the ideas of Plato and Aristotle became bedrock, upon which the ideological superstructure of the medieval Church was built.

We are now in a position to neatly summarise the contrasting views of the Atomists and Aristotelians.

A True Story: Greek alien fiction

And yet, from those ancient days comes a work of speculative fiction that is just bristling with irreverent and scandalous ideas about extraterrestrial life. The tale is *A True Story*, and the author of the work is Lucian, an Assyrian rhetorician and satirist who wrote in Greek. Noted for his witty and scoffing nature, Lucian was one of the first novelists in the West, and *A True Story* is the

Table 1.2. *A comparison of the cosmologies of the Atomists and Aristotelians*[66]

Cosmos of Atomists	Cosmos of Aristotelians
Homogeneous	Hierarchical
Infinite	Finite
Random	Ruled by design
Purposeless	Teleological
Atheistic	Theistic
Cosmos is mutable and earthly	Cosmos is immutable and perfect
Vacua exist	Vacua do not exist
Plurality of worlds	Unitary Earth

earliest known fiction about space travel and alien life. His tale contains themes of interplanetary imperialism and warfare, predating that kind of space opera by many centuries.

Lucian's is a satirical narrative work, written in prose, in which he parodies the kind of fantastic tales told by Homer in his *Odyssey*, and other more feeble fantasies, popular in Lucian's time. *A True Story* is a remarkable work. It anticipates many features of modern science fiction, such as the fictional travelogues to Venus and the Moon, contact with extraterrestrial life-forms, and wars between planets. And it was written seventeen centuries before the likes of H. G. Wells and Jules Verne practically invented the genre in more modern times. *A True Story* is, in short, the earliest work of astrobiological science fiction.

The tale begins as Lucian and a company of adventuring heroes sail westward through the Straits of Gibraltar. They wish to explore the lands and peoples beyond the Ocean, but are blown off course by a strong wind. In due course, they are lifted up by a giant waterspout, a rather ingenious form of lift-off, and set upon the Moon:

> For seven days and seven nights we sailed the air, and on the eighth day we saw a great country in it, resembling an island, bright and round and shining with a great light. Running in there and anchoring, we went ashore, and on investigating found that the land was inhabited and cultivated. By day nothing was in sight from the place, but as night came on we began to see many other islands hard by, some larger, some smaller, and they were like fire in color. We also saw another country below, with cities in it and rivers and seas and forests and mountains. This we inferred to be our own world.[58]

This is an interesting passage, as Lucian skilfully plays down this initial encounter with the Moon and other celestial bodies. The wealth of extraterrestrial

Wonderheden LVCIANI. 15

CAPITEL IIII.

Lucianus treckt ten velde met het volck van de Mane tegen het volck van de Sonne. Beschrijvinghe van beyde de leghers.

DOen vraeghden wy den Coninck / wie syn vyanden waren / ende waerom dese oorloghe gheresen was. Doen seyde hy: Den Coninck van de Sonne / de welcke alsoo wel bewoont is als de Mane / Phaeton ghenaemt / heeft langhe ende menighen tyt oorloghe ghevoert teghen ons: welcke oorloghe is ghesproten om de reden die ghylieden hooren sult: In voorleden tyde soo hadde ick vergadert alle de arme lieden die in myn rycke vander Mane waren: de welcke ick gheerne ghestelt hadde om te woonen in't landt vande dagh-sterre / Lucifer ghenaemt / welck doen

Figure 3 Lucian of Samosata, *A True Story*, Dutch edition, 1647.

wonders still to come, Lucian first focuses on our sense of familiarity, with reference to an 'island' and to a land that is 'inhabited and cultivated'. This emphasis serves to show that our world, from the narrator's point of view on the Moon, looks just like another land 'below', and is otherwise identical to any other extraterrestrial planet.

Once on the Moon, they find a war raging, between the king of the Moon and the king of the Sun, over the colonisation of the Morning Star. The armies that wage this war are staffed with the most exotic of militia. There are cloud centaurs, stalk-and-mushroom men, and acorn-dogs, 'dog-faced men fighting on winged acorns'.[59] Tellingly, the Moon, Sun, stars, and planets are depicted as realistic backdrops, each with its own distinct habitat and peoples. Presently, the war is won by the Sun, and the Moon is clouded over. But not before more details of the Moon are unveiled: there are no Moon women, children being grown inside the Moon men.

This tale of Lucian's is remarkably replete with alien life, and is an ancient example of 'first encounter' fiction, stories in which humans come across extraterrestrial life. Indeed, one section of the story wittily describes men being seduced, and then incorporated in a Borg-like manner,[60] by an alien life-form. Another section seems to suggest that the Moon people possess a well and a mirror, by which doings on Earth may be heard or watched respectively.

Everything in this work of Lucian's is touched with a sense of wonder. True, much of this is due to satire and the way in which Lucian continually expresses amazement and wonder at each novel experience in his *voyage extraordinaire*. This interest in the strange and bizarre, and the mere thrill of it, is one reason for Lucian's later popularity with creative Renaissance thinkers. The two worlds share a thirst for the shock of the new. It also suggests why, uniquely among the ancients, Lucian seems so utterly contemporary. Like the days of Columbus and Copernicus and the paintings of Hieronymus Bosch, the imaginative horizons of Lucian's world seem transcended time and again by the prospect of new discoveries.

To what extent is *A True Story* a text about the science of pluralism? Though the author tells the reader in the Preface that the work is meant only as intellectual amusement, it is nonetheless useful to consider Lucian's narrative as an indicative text: that is, as a measure of significant changes in attitude and imagination due to Atomist philosophical ideas about life beyond the Earth.

Clearly the sciences inspire Lucian's creative and imaginative work. Astronomy is prominent in the account of the trip to the Moon and beyond, and although his cosmology is unremarkable, it does tend to the Atomist, as it portrays the heavenly bodies as other worlds on a par with our own. Lucian's geography is also worldly, in the sense that it mocks allegedly factual contemporary accounts that are outrageously fictive. It is the same with natural history. Rather than reduce his attitude to the usual Greek anthropocentrism, typical of the age, Lucian shows sensitivity in his descriptions of life like insects, spiders, molluscs, and fishes.[61]

In such ways *A True Story* shows how the physical sciences, and especially Atomist pluralism, had affected his imagination. They have influenced ways of thinking, as well as content. Indeed, there has been significant travel since the days of Homer, whose epics the *Iliad* and the *Odyssey* date from around the eighth century BC. The worldview implied by the astronomical, biological, and anthropological journey of Lucian's narrator is of an entirely different intellectual order than the overwhelmingly mythical landscape negotiated by Odysseus in the *Odyssey*.[62]

There are other ways in which *A True Story* is revolutionary. With this work, Lucian takes contemporary philosophical thought and uses it to picture alternative worlds in a way that makes his readers look at things afresh. They become somewhat dislocated from their routine way of thought, and are forced to contemplate other worlds and other places. Like many who were to follow, Lucian seduces the reader into speculating on the habits of mind in the real world. Little wonder then, that Lucian's story is regarded by many as being the first true work of science fiction, the genre renowned for its depiction of alternative worlds, radically unlike our own, but somewhat similar in terms of scientific knowledge.

The most significantly 'astrobiological' aspect of Lucian's story is his vivid portrayal of the other planets, especially his Moon world. He begins by describing the war between the Moon and the Sun as if they were expansive, mongrel empires of the Hellenistic world. Here we find the weapons of war, the considered tactics and strategies, and even scientific detail of a battle wound. Then, we get the anthropology and biology of this lunar world. The native people are described in terms of their clothing, their diet, and substantial detail of their physiology, including their distinct alimentary, urinary, excretory, and reproductive systems. Indeed, Lucian's refinement is such that he goes beyond simple ideas. Witness his gender portrait of the lunar people, who he does not describe as 'all-male', but says rather that they do not even have a word for 'women'.

Lucian's lunar world is a land of giants. Birds and plants, spiders and ants, mosquitoes and cloud centaurs – all are of monstrous proportions. Such beasts of mammoth dimension are emphasised further by the sheer number and variety of alien creatures and sub-human forms that are paraded before us. Lucian uses this giganticism as a distancing effect. When faced with ants employed as battle mounts, vegetables for boats, and nuts and seeds for weapons and armour, the reader views human behaviour with detachment and irony, when it is projected onto weirdly alien creatures like the Moonites. In this way, the Moon world is presented as a parallel Earth, with all the same intricacies as the home planet. Like the fiction that was to follow in the wake of the scientific revolution centuries later, Lucian's text begins with a fictional hypothesis, but grows out of an extrapolation of a scientific framework. The other alternative worlds are not portrayed in such detail, but are nonetheless philosophically remarkable.

The most striking of these is the world within a whale. Lucian's picture of this surreal landscape contrasts not only with the real world, but also with the extraterrestrial one preceding it. Inside the whale we encounter two Greeks, living apart from other men in a Crusoesque fashion, who have managed to impose themselves, through viticultural technology, on the 'natural' environment. This minimalist and microcosmic model of culture is further telescoped when the narrator taps into the castaways' innate proclivity for war by joining forces and totally exterminating various fish and mollusc races in the belly of the whale. Whether a realm is open or closed, Lucian seems to be saying, the world is fit for human exploration, and exploitation, through the application of knowledge.

Some of the alien worlds Lucian explores are partial and cultural, rather than total and physical. One such alternative world is Elysium, the Isle of the Blessed. Only the dead inhabit this utopian fantasy, as the living are barred from entry. With a most mischievous sense of humour Lucian selects, as members of this society, famous dead people about whom some controversy has raged, for instance Pythagoras and Empedocles, and has them answer their critics. So, in this witty section of the travelogue, Lucian is able to comment on many fields of knowledge, including the philosophy of several schools.

Now, an almost endless procession of alien races seems to pour past. The doughty travellers reach the underworld of Tartarus, the first dystopia in science fiction. Here they briefly meet alternative human cultures in the form of pumpkin-pirates, nut-sailors, and dolphin-riders. Next come groups of alternative worlds that are consciously nonhuman in design. Landscapes composed entirely of wine or milk products suggest an alternative ecology. Nonhuman worlds inhabited by alien races are rational, rather than mythic, as they are scrutinised in both nature and anthropology. Witness the worlds of the vine-nymphs, the bullheads, a race of minotaurs, the asslegs, a cannibalistic race of shape-changing women, and a race of feuding giants on floating islands. All extraterrestrials are portrayed as distinct from, but analogous to, men.

Lucian's 'Lamptown' creatures are even more remarkable. As the passage quoted at the top of this chapter shows, Lucian seems to have created an alternative and civilised alien life-form, which hints at the robots and AIs of twentieth-century film and fiction. The product of human technology, not only do these lamp-creatures occupy 'a city [that] lies in the air midway between the Pleiades and the Hyades',[63] but Lucian even has a conversation with his own lamp, 'I spoke to him and enquired how things were at home, and he told me all about them'.[64]

Cloudcuckooland comes next. A passing reference to this imaginary land in the clouds, culled from Aristophanes' comedy *The Birds*, is made before our narrator sails on. Finally, there remains the 'other world' across the western ocean, the initial destination of the journey, but one never reached. The story ends with a promise that the next volume would feature adventures in this other world beyond the ocean, as if Lucian is suggesting he can go on generating these extraterrestrial worlds without limit, and the world of the imagination – unlike the real world of late antiquity – remains forever open.

Lucian greatly admired the works of Epicurus. In one of his most witty satires, the one against the Greek mystic and oracle Alexander of Abonoteichus, Lucian vents his spleen, describing Alexander as a swindler, and of engaging, along with his followers, in various forms of thuggery. Alexander's hatred of the Epicureans, and his burning of a book by Epicurus, explain the strength of Lucian's venomous attack:

> What blessings that book creates for its readers and what peace, tranquillity, and freedom it engenders in them, liberating them as it does from terrors and apparitions and portents, from vain hopes and extravagant cravings, developing in them intelligence and truth, and truly purifying their understanding, not with torches and squills and that sort of foolery, but with straight thinking, truthfulness and frankness.[65]

As an admirer of Epicurus, and like other Atomist philosophers, Lucian was a pluralist of the imagination. He had a questioning mind that traces back to the old Ionians, a curiosity about the wonderful variety in nature and human society alike. In contrast to the idealists, Lucian was a thorough sceptic. He was very aware that in many ways the old discarded myths were re-emerging in the disguised form of some of the new philosophies, and the degeneration of thought into pseudo-science. Lucian's tale is the most ancient example of what has come to be known as the intellectual nonconformism of science fiction. And the wealth of estranged worlds of *A True Story* is the shape of things to come.

Notes

1 Lucian, *A True Story*, trans. A.M. Harmon (1913), parallel English and Greek, Loeb Classical Library No. 14, pp. 247–357, Lucian Vol. 1
2 Isaacs, L. (1977) *Darwin to Double Helix: The Biological Theme in Science Fiction*, London and Boston, p. 6
3 Whitehead, A.N. (1953) *Science and the Modern World*, Cambridge
4 Koestler, A. (1959) *The Sleepwalkers: A History of Man's Changing Vision of the Universe*, London
5 Bernal, J.D. (1965) *Science in History, Vol. I*, London
6 Petron, one of the early Pythagoreans, is alleged to have held that there were just a hundred and eighty-three worlds arranged in a triangle.
7 Simplicius, *Commentary on Aristotle's Physics*, 1121, 5–9
8 ibid.
9 Cicero, *On the Nature of the Gods* (I, 10, 25)
10 Hippolytus, *Refutation of all Heresies*, I, 6
11 Bernal, J.D. (1965) *Science in History, Vol. I*, London
12 *The School of Athens* is one of the most famous paintings by the Italian Renaissance artist Raphael. The picture has long been seen as Raphael's masterpiece and the perfect embodiment of the classical spirit of the High Renaissance.
13 Bernal, J.D. (1965) *Science in History, Vol. I*, London
14 ibid.
15 Koestler, A. (1959) *The Sleepwalkers: A History of Man's Changing Vision of the Universe*, London

16 Farrington, B. (1953) *Greek Science*, London
17 Bernal, J. D. (1965) *Science in History, Vol. I*, London
18 Plato, incidentally, derided the philosopher Protagoras, and other sophists of the fifth century, for the fact that they took fees for their teaching. Plato, who was rich enough not to bother, mocked them for lowering the status of the sage.
19 Guthke, K. S. (1990) *The Last Frontier: Imagining Other Worlds from the Copernican Revolution to Modern Science Fiction*, New York
20 Quoted in Bernal, J. D. (1965) *Science in History, Vol. I*, London
21 Censorinus, *De Die Natali*, IV, 7
22 Mason, S. F. (1953) *A History of the Sciences*, London
23 Diels, H. A. (1952) *The Fragments of the PreSocratics*, Berlin, B54
24 Bernal, J. D. (1965) *Science in History, Vol. I*, London
25 A doxographer is a biographer of past philosophers.
26 Diogenes Laërtius Book III.20.53
27 Levenson, J. (1991) *Circa 1492: Art in the Age of Exploration*, New Haven
28 Russell, B. (1972) *A History of Western Philosophy*, London
29 Gangopadhyaya, M. (1981) *Indian Atomism: History and Sources*, Atlantic Highlands, New Jersey
30 McEvilley, T. (2002) *The Shape of Ancient Thought: Comparative Studies in Greek and Indian Philosophies*, New York
31 Teresi, D. (2003) *Lost Discoveries: The Ancient Roots of Modern Science*, London
32 Cicero, *de Finibus*, v. 19, Strabo, xvi
33 Diogenes Laërtius, ix, 34
34 Cicero, *de Finibus*, v. 19, Strabo, xvi
35 Clement of Alexandria, *Stromata*, i
36 Diogenes Laërtius, ix, 72
37 Stevenson, J. (2005) *The Complete Idiot's Guide to Philosophy*, Indianapolis
38 Epicurus, trans. C. Bailey (1957) Epicurus to Herodotus, in *The Stoic and Epicurean Philosophers*, ed. W. J. Oates, New York
39 ibid.
40 Interestingly, the atheism of the Epicureans is legend. The word 'Apiqoros' appears in ancient Judaism, referring to either those who lack religion or are seemingly atheistic. The origin of the word is contentious, but is likely to refer to Epicurean beliefs, although it came to imply any philosophy lacking a god. On occasion it is even used to describe heretic ideas or the heretics themselves. The word is still used in modern Hebrew.
41 Epicurus, trans. C. Bailey (1957) Epicurus to Herodotus, in *The Stoic and Epicurean Philosophers*, ed. W. J. Oates, New York
42 As quoted from Simplicius in Cornford, F. M. (1964) Innumerable Worlds in the Pre-Socratic Philosophy, *Classical Quarterly*, **28**, 13
43 Titus Lucretius Carus, *On the Nature of the Universe*, trans. R. E. Latham (1975) Baltimore, pp. 90–3
44 ibid.
45 ibid.
46 ibid.
47 ibid.
48 Dreyer, J. L. E. (1906) *History of Planetary Systems from Thales to Kepler*, Cambridge

49 Heath, T.L. (1932) *De facie in orbe lunae*, Chapter 6 of *Greek Astronomy*
50 Bernal, J.D. (1965) *Science in History, Vol. I*, London
51 *The Republic of Plato*, book VII, trans. T. Taylor
52 Grundy, G.B. (1956) article on Greece, *Encyclopedia Britannica*, x-780c
53 Aristotle, *On the Heavens*, trans. J.L. Stocks, in *The Basic Works of Aristotle*, ed. R. McKeon, New York, p. 409
54 ibid.
55 ibid.
56 ibid.
57 Aristotle, *Metaphysics*, trans. W.D. Ross, in *The Basic Works of Aristotle*, ed. R. McKeon, New York, p. 884
58 Lucian, *A True Story*, trans. A. M. Harmon (1913), parallel English and Greek, Loeb Classical Library No. 14, pp. 247–357, Lucian Vol. 1
59 ibid.
60 A fictional race of cyborgs in the Star Trek universe
61 See 'Lucianic Natural History' in *Classical Studies Presented to Ben Edwin Perry* (Urbana, Ill., 1969; Illinois Studies in Language and Literature, vol. 58), pp. 15–26, in which Jerry Standard demonstrates convincingly that Lucian was a competent observer of nature and was acutely sensitive to details of anatomy and animal behaviour, so that even his most wildly fictitious and imaginary plants and animals are enlivened by quasi-realistic touches. Witness the narrator's dissection and inspection of one of the fishes native to the river of wine on Dionysus' island.
62 See Bompaire, J. *Lucien Écrivain* (Paris, 1958), p. 659 for a lengthy catalogue of parallels with the Odyssey.
63 Lucian, *A True Story*, trans. A. M. Harmon (1913), parallel English and Greek, Loeb Classical Library No. 14, pp. 247–357, Lucian Vol. 1
64 ibid.
65 Richter, D.S. (2005) *Lives and Afterlives of Lucian of Samosata*, Arion 13.1:75–100.
66 After Crowe, M.J. (2008) *The Extraterrestrial Life Debate: Antiquity to 1915*, Notre Dame, p. 13

2

The world turned upside down: Copernicanism and the voyages of discovery

Whatever is born on the land or moves about on the land attains a monstrous size. Growth is very rapid. Everything has a short life, since it develops such an immensely massive body.

The Privolvans have no fixed abode, no established domicile. In the course of one of their days they roam in crowds over their whole sphere, each according to his own nature; some use their legs, which far surpass those of our camels; some resort to wings; and some follow the receding water in boats; or if a delay of several more days is necessary, then they crawl into caves. Most of them are divers; all of them, since they live naturally, draw their breath very slowly; hence under water they stay down on the bottom, helping nature with art. For in those very deep layers of the water, they say, the cold persists while the waves on top are heated up by the sun; whatever clings to the surface is boiled out by the sun at noon, and becomes food for the advancing hordes of wandering inhabitants.

For in general the Subvolvan hemisphere is comparable to our cantons, towns, and gardens; the Privolvan, to our open country, forests, and deserts. Those for whom breathing is more essential introduce the hot water into the caves through a narrow channel in order that it may flow a long time to reach the interior and gradually cool off. There they shut themselves up for the greater part of the day, using the water for drink; when evening comes, they go out looking for food.

In plants, the rind; in animals, the skin, or whatever replaces it, takes up the major portion of their bodily mass; it is spongy and porous. If anything is exposed during the day, it becomes hard on the top and scorched; when evening comes, its husk drops off. Things born in the ground – they are sparse on the ridges of the mountains – generally begin and end their lives on the same day, with new generations springing up daily.

In general, the serpentine nature is predominant. For in a wonderful manner they expose themselves to the sun at noon as if for pleasure; yet they do so nowhere but behind the mouths of the caves to make sure that they may retreat safely and swiftly.

Johannes Kepler, *Somnium*, trans. by E. Rosen[1]

The voyages of discovery

It all began with ships.

Two Chinese inventions, the compass and the sternpost rudder, were to have a global effect at sea. Long voyages became viable. The seas were thrown open. Open to exploration. Open to piracy. Open to a colossal expansion in trade. And open to war.

The need for better navigation had profound consequences for science. An open ocean meant more accuracy: better observations, better charts, and better instruments. So open-sea navigation raised the need for a more predictive astronomy, a brand-new quantitative geography, and the desire for devices that could be used on board ship, as well as on land.

The obsession with longitude began. Deep-sea navigation also meant an urgent need to know where on Earth you were. Mariners for most of history had struggled with precise longitude. Latitude was easy enough; observing or predicting the positions of the Sun, Moon, planets, and stars would easily nail it. Mariner's tools such as the quadrant or astrolabe were used, reading off the inclination of the Sun or charted stars.

Longitude presented no such manifest means of study. All the great astronomers of the day tried their hand. The need for compasses and other devices for navigation ignited a new skilled industry, especially that of the card and dial makers.[2] The ensuing influence on science was huge. It set new standards for precision, and many prominent scientists were also instrument makers, including Newton, Watt, and, of course, Galileo.

The great European sea voyages started with the Portuguese explorers around 1415, and opened up the planet to capitalist enterprise. The voyages were the fruit of the first conscious use of astronomy and geography in the pay of glory and profit. For fledging empires soon realised they were able to exert global control, based on knowledge of territory: knowing *where* you were, knowing *what* you owned. And so astronomy, navigation, and mapping became even more important to trade. German scholars, such as Peurbach and his pupil Regiomontanus of Nurnberg, later assisted by Albrecht Dürer, refined the application of astronomy to navigation. A drive was initiated for ephemerides plain and precise enough for the ocean navigators' needs.

The ship was of prime importance. As all the major trade routes were wet, commercial control very much depended on the speed and consistency with which long-range voyages took place along those wet routes. The discovery of the Americas and the New World was the impetus for a more aggressive approach to nature. The quest for longitude and dominion in that New World promoted instrument making in science. And one of the first instruments of discovery was the ship itself.

The telescope too became a kind of ship. Consider this: pioneers of the spyglass took themselves, and their contemporary world as onlookers, to a place almost no one had ever imagined. But, like the destinations of a ship, the destinations of the telescope were also public knowledge. If you disbelieved what the likes of Galileo claimed, you could take a look for yourself.

No one has put a better case for the ship as an instrument of discovery than twentieth-century German playwright Bertolt Brecht, in his greatest play, *Life of Galileo*. In the very first scene, Brecht has the Tuscan astronomer declare from his 'humble study in Padua',

> For two thousand years men have believed that the Sun and the stars of heaven revolve around them . . . The cities are narrow and so are men's minds. Superstition and plague. But now we say: because it is so, it will not remain so. I like to think that it all began with ships. Ever since men could remember they crept only along the coasts; then suddenly they left the coasts and sped straight out across the seas.
>
> On our old continent a rumour started: there are new continents! And since our ships have been sailing to them the word has gone round all the laughing continents that the vast, dreaded ocean is just a little pond. And a great desire has arisen to fathom the causes of all things . . . For where belief has prevailed for a thousand years, doubt now prevails. All the world says: yes, that's written in books but now let us see for ourselves.[3]

And so it was that theory and custom met at the Court of Prince Henry the Navigator at Sagres, at the south-western tip of the European mainland. There, Italian, German, and Moorish scholars planned voyages with seasoned Portuguese and Spanish sea captains. The Turkish throttlehold on eastern trade raised the compelling idea of venturing into the Indian Ocean by some way other than the Red Sea. Scholars and strategists considered two promising alternative routes. The first, to round Africa, was profitably realised by the Portuguese in 1488, though Vasco da Gama did not reach India until 1497.

The second route was nothing short of revolution. The proposal, made by astronomers and geographers such as the Florentine Paolo dal Pozzo Toscanelli, was to head west over the uncharted ocean, in search of China at the other side of the world. Now to imagine such a hypothesis is one thing. But to sail straight

SCIENTIFIC AMERICAN

A WEEKLY JOURNAL OF PRACTICAL INFORMATION, ART, SCIENCE, MECHANICS, CHEMISTRY, AND MANUFACTURES.

NEW YORK, OCTOBER 25, 1890.

M. PALACIO'S DESIGN FOR A COLOSSAL MONUMENT IN MEMORY OF CHRISTOPHER COLUMBUS.

Figure 4 Colossal monument to Christopher Columbus, designed by Bilboan architect M Alberto de Palacio for the Universal Exposition of 1892. The sphere would have had a diameter of nearly a thousand feet, equal to the height of the Eiffel Tower (*Scientific American*, vol. 63, no. 17, 25 October 1890).

out to sea is quite another. In the popular mind, a host of possible fates might befall such explorers. They might sail on without end, or they might plunge off the edge of the world. Few indeed thought there would be an entire continent in the way.

The shipman who finally took the risk was the supreme navigator and adventurer, Christopher Columbus. Though he was far from being a scientist, Columbus had a brilliant vision.[4] By seafaring west, new worlds or even 'a new heaven and a new earth' may be discovered. And it was this dream, part spiritual, part scientific, that finally seduced the venture capitalists.

There was a deep distinction between the successive voyages of the Portuguese round Africa, and of Columbus risking all to set sail directly across the sea.

The Portuguese depended on a gradual refinement of the customised route. Columbus' proposal was scientific: it offered a revolutionary break with such tradition. True, there may have been mystical motives. But the finance Columbus received for his voyage was an investment in science – the promise of booty plundered from the proof of a theory. It is this spirit of Columbus that captures the mood of the age. The unearthing of the New World made the ancient Greek world seem provincial.[5] And early modern philosophers and authors alike were won over by discoveries of which the classical world had never dreamt. Jean Fernel,[6] physician to the King of France, expressed the new spirit in his *Dialogue* in about 1530:

> it seems good for philosophers to move to fresh ways and systems; good for them to allow neither the voice of the detractor, nor the weight of ancient culture, nor the fullness of authority, to deter those who would declare their own views. In that way each age produces its own crop of new authors and new arts.
>
> The continent of the West, the so-called New World, unknown to our forefathers, has in great part become known. In all this, and in what pertains to astronomy, Plato, Aristotle, and the old philosophers made progress, and Ptolemy added a great deal more. Yet, were one of them to return today, he would find geography changed past recognition. A new globe has been given us by the navigators of our time.[7]

Finally, a notion began to stir in the liberated imagination: if new worlds were being discovered on Earth, why not also in space?

The medieval cosmos

So the movement of the stars now had a cash value. But there was a problem in heaven.

If Aristotle had been the last great cosmologist of antiquity, Ptolemy was its last great astronomer.[8] And the writings of these two ancient philosophers cast a long shadow over medieval astronomy, eclipsing much thought in the Middle Ages on the question of the cosmos. Indeed, and as we shall see, Copernicus and Galileo might as well have been their immediate successors. As, in the one and a half millennia that yawn between the death of Ptolemy and the birth of Galileo, no lasting change had been made in astronomy.

The *Almagest*, Ptolemy's magnum opus, was the sole surviving ancient treatise on astronomy of any note. Written in the second century AD, the tome comprised thirteen books. It was antiquity's 'state-of-the-universe' report on Aristotelian cosmology – a guide to the intricate motions of the stars and planets. Within its covers, Ptolemy marshalled a model based on the work of his predecessors, spanning more than eight hundred years. And yet practising astronomers had known, for centuries after, that the models he created were divorced from actual observation.

Astronomy, as a creed of the idealists, had become an abstract sky geometry, divorced from observation and experiment. And Ptolemy's model, first catalogued in AD 150, had become the new standard bearer of ancient astronomy. When the Christian West recovered some of the ancient learning from the Arabs, it was unsurprisingly in Arabic translation. For this reason, Ptolemy's great work is not known by its Greek name at all, but by a sawn-off Arabic title it got from a ninth-century Moslem translator who called it *Almagest*, literally 'The Great Book'.[9]

Ptolemy's book had held such sway that his system was still dominant when English poet John Milton wrote his epic saga *Paradise Lost*, in the seventeenth century. The apparent aim of the work was to 'justify the ways of God to men',[10] but Milton assessed the Ptolemaic model with some ridicule:

> From man or angel the great Architect
> Did wisely to conceal, and not divulge,
> His secret to be scanned by them who ought
> Rather admire; or, if they list to try
> Conjecture, he his fabric of the Heavens
> Hath left to their disputes, perhaps to move
> His laughter at their quaint opinions wide
> Hereafter, when they come to model Heaven
> And calculate the stars, how they will wield
> The mighty frame, how build, unbuild, contrive
> To save appearances, how gird the sphere
> With centric and eccentric scribbled o'er,
> Cycle and epicycle, orb in orb.

But maybe the final word on Ptolemy's 'quaint opinions wide' goes to the thirteenth-century King of Castile, Alfonso X. The king (known also by his nicknames of 'el Sabio', 'the Wise', and 'el Astrólogo', 'the Astronomer') uttered the view, when first presented with the Ptolemaic system, 'If the Lord Almighty had consulted me before embarking upon Creation, I should have recommended something simpler'.[11]

No wonder there was little talk of the alien after Lucian.

Half a millennium stood between Aristotle and Ptolemy. But when their works became Latinised, they seemed to fuse. For starved men emerging out of the Dark Ages, the ancient wisdom was breathtaking in its scope, and dazzling in its brilliance. Those who pored over their words were captivated by their majestic cosmology. So began the task of assimilating this ancient and spellbinding knowledge. Ptolemy's system had neither heir nor rival. In Europe and the Arab world, the great majority of the technical astronomy done during the preceding fourteen centuries was devoted to the calculation of tables and the design of instruments, all in honour of *Almagest*.

So Ptolemy was the word. And the word became almanac and horoscope, planetaria and ephemerides. To the modern eye, it is maybe too easy to denounce medieval astronomers for their lack of critical thought. Is it not obvious, to the meticulous watcher of the skies, that the heavens revolve around the Sun, and not the Earth? One might wonder what was so hard about ridding astronomy of the geocentric mindset: the terrestrial chauvinism of central Earth, immutable heavens, and uniform circular motion of the planets. After all, these conceits led nowhere, other than to abstract babble such as Ptolemy's model.

But the reality was not so prosaic. Consider an apparently simple puzzle: to prove heliocentrism, with the evidence available to Greek astronomers. Or, to put it another way, to show the Solar System was Sun-centred, with the facts at hand in the medieval period before the telescope. The circumstances had remained essentially the same, for the technology available to explore the sky had changed little between those times.

Even to the most meticulous observer, the entire cosmos – Moon, planets, stars, and the rest – seems to wheel about the central Earth, all rising in the east, and setting in the west. The Earth-centred system accounts for this daily parade in a simple enough manner. Some of the more irregular cosmic events, however, and even the odd spectacular episode are also explained away using the geocentric model. Though the queer motion of the planets seemed problematic, lunar and solar eclipses incredible, and regular changes in a body's brightness troublesome, Ptolemy's canny system was enough to appease the doubters, and explain away the glitches.

As for other heavenly happenings, the likes of ominous comets or the seldom-spotted supernovae, such phenomena could also be excused. Aristotle was adamant that such occurrences were merely disturbances in our atmosphere – the sublunary sphere from the Earth to the Moon was the only realm allowed any change in his cosmological scheme of things.

So from Plato and Aristotle through Ptolemy and beyond, few doubted the terrestrial chauvinism of the geocentric model. If the abstractions did not precisely fit observation, reason and common experience condoned it. Philosophers and journeymen, poets and paupers, noblemen and kings, all spoke of the universe much as the ancient Greeks had described it. And laced into this conceit, since only Earth came accompanied with death and decay, was the idea that this was the sole place in the entire universe where life resided.

The medieval Throne of God and the alien

Consider Dante Alighieri's epic poem, the *Divine Comedy*, beautifully realised in Figure 5 in a painting called *Dante and the Three Kingdoms*, by Domenico di Michelino. Dante's famous work is a fictional account of the poet's epic journey through the fourteenth-century Christian cosmos.

The journey begins on the surface of the spherical Earth. Dante then descends into the Earth, through the nine circles of Hell, which mirror the nine spheres of heaven in Aristotle's cosmology, seen in Michelino's picture. Dante travels through the sixth circle of hell, whose contents include the flaming tombs of the ungodly Epicureans. Presently, Dante reaches his first destination, the most corrupt of all realms, the squalid centre of the universe, locus of the Devil and his legions.[12]

Figure 5 Domenico di Michelino's fresco *Dante and the Three Kingdoms* (1465), Museo d'Opera del Duomo, Florence. The painting shows Dante, holding his book aloft, and surrounded by the three realms: Purgatory behind him, the heavenly City at his left, and to his right a procession of tortured souls descend into the circles of Hell, to which Dante gestures.

The poet then re-emerges at the other side of the globe, where he comes upon the Mount of Purgatory, shown in Michelino's painting with its base on the Earth and its peak, complete with Adam and Eve in the Garden of Eden, reaching high into the heavens above. Passing through purgatory, Dante now wings his way through the aerial regions, now passing through the terrestrial spheres of air and fire, then flying through each celestial sphere, speaking with the spirits that inhabit each of them. Ultimately, he approaches the Primum Mobile, the first moved thing, and the sphere that gives motion to the rest of the realm. Finally, Dante beholds the last sphere, the Empyrean, God's Throne.

The universe of Dante's *Divine Comedy*, and Michelino's portrait, is Aristotle's, tailored to the medieval Holy Church, and God himself. The only life encountered on this journey through heaven and hell is either terrestrial, or divine. No other single soul stirs beyond the sphere of the Moon. No one lives in the Sun, no Martians dwell on the red planet, gone are Lucian's flights of lunar fancy.

The alien is nowhere. And for good reason. It is a universe malformed by Christian symbolism. By his use of allegory, Dante implies that the medieval universe could have no other structure than the Aristotelian, and in Aristotle's picture the Earth was unique. In Dante's powerful poem, a masterpiece of world literature, Aristotle's universe of spheres mirrors man's hope and fate.[13] Both bodily and spiritually, Man sits midway. His pivotal position in the hierarchical chain is halfway between the inert clay of the Earth's core and the divine spirit of the Empyrean. The rest of the cosmos is made of either matter or spirit. But uniquely, Man is made of both, body and soul.

Man's place is also transitional. He lives on Earth, in filth and insecurity, close to Hell. But he is at all times and in all places under the all-seeing eye of God, and with full knowledge of the heavenly escape above. Man's duality and place in the great scheme of things helped impose the dramatic choice that faced all Christians in the medieval world: whether to follow his base and human nature down to its natural place in Hell, or to engage with his spirit, and follow his soul up through the celestial spheres to God.

Such a stark picture makes it clear why any earthly life beyond this world would have been untenable. The whole edifice would come crumbling down at the mere thought of men on Mars. This epic poem, *The Divine Comedy*, is 'the vastest of all themes, the theme of human sin and salvation, is adjusted to the great plan of the universe'.[14] A great adjustment had been achieved. The Western world lay at the feet of the medieval Holy Church. Any alteration in the great plan of Aristotle's universe was bound to corrupt the drama of Christian life and death.

To shift the Earth was to sever the continuous chain of created being, to undermine this Christian drama, and to move the Throne of God himself. And yet this is what now happened.

The Copernican revolution

A new economy was born, and with it, the Renaissance.

Open-sea navigation led to booty, and plenty of it. The old and expensive dry trade routes were superseded by fast and cheaper wet ones, the most spectacular being with America. The obsession with plundering more profit fed back into the drive for shipbuilding and navigation. Progress in print production meant that the dissemination of subjects such as agriculture, cooking, and trade was far easier, and so the process accelerated.[15]

The pirates that preyed upon the high seas, an echo of rival trade and colonisation attempts by European powers, often sought a surprising booty. If a sea raid proved successful, the boarding pirates would head straight for the hold. For rather than gold, silver, or pieces of eight, the most precious cargo a ship possessed was its maps and chronometer. Indeed, some cartographers would knowingly include errors on their maps, to mislead the uninitiated should the map get into the hands of the wrong kind of pirate.

And key to all this was the spirit of revolution in the air. In science, and society, a conscious vanguard of merchants, scholars, and artists set about the task of constructing a new culture, capitalist in its economy, classical in its art, and scientific in its approach to nature.[16] Initially, the plan had been to reconstruct the classical world, returning to Plato and Aristotle, Democritus and Archimedes. But soon the discovery of the New World made that ancient world seem unfashionable.

Now, the fusion of knowledge and action was explosive.

The main intellectual mission of the Renaissance was to rediscover and master nature. Astronomy became the acid test of progress, a grindstone against which the new science was sharpened. And its greatest achievement was a cosmology with the Sun at the centre. Nicholas Copernicus' clear and detailed explanation of a rotating Earth in a Sun-centred system was a decidedly descriptive astronomy, somewhat devoid of observation and experiment. Nonetheless, he was possessed of the new critical spirit.

The picture Copernicus painted in *De Revolutionibus Orbium Coelestium* (*On the Revolutions of the Celestial Orbs*, 1543) was clear enough for all to see. It was a system of the new lay society. True, it was founded, at least in part, on the classical learning. But there was a huge change. The new cosmos could be seen and experienced for itself. At the start, the political impact of this new learning was not evident. Only with Galileo's evidence through the telescope did the Church take fright and tried, too late, to shut it out.[17]

The system of Copernicus was nothing new, of course. In the third century BC, Aristarchus had not merely imagined a rotating Earth, he had also set out the first heliocentric cosmos. His system had been dormant ever since, a paradoxically alternative and absurd cosmology, one that suggested it was the Earth that moved, even though it was patently the Sun, Moon, and stars that could be seen to do so.

Armed and inspired by newly edited classical texts, a capable humanist could balance one Greek authority with another: Democritus against Plato, Epicurus with Aristotle, Aristarchus against Ptolemy. Equipped with a strong aesthetic sense, and bags of bravery, such a humanist might even dare to centre a new universe about the Sun, and worry later about shifting the Throne of God.

And that is just what Copernicus did.

He had the courage to confront common sense, enough nous in science to make the Sun star of his system, and, as a Renaissance humanist, enough motivation to help bring the whole edifice of ancient thought plummeting down. Copernicus had been born in Toruń, Poland, in 1473, a mere generation into the Renaissance. Born of the patrician class, his mother was from a family of rich merchants, his father a wealthy and respected copper trader of that city. Indeed, the Polish version of their surname, *Kopernik*, means 'one who works with copper'. Nicholas Copernicus was a man of his time. Witness his Renaissance training in Italy: cultured in astronomy at Bologna, medicine at Padua, and law at Ferrara. He spent most of his adult life as a canon in Frauenberg, a cathedral town situated in a disputed territory. With the kingdom of Poland on one side, and the Teutonic knights on the other, Copernicus was often working with the offices of war.

Astronomy was both refuge and fascination. Copernicus seems to have given his entire private life over to the cause of building his own cosmic vision. For the last fifteen years of his life, he refined his heliocentric theory. Frightened of the social, political, and religious ramifications his theory would arouse, he did not set out his model in its final form until he lay on his deathbed. *On the Revolutions of the Celestial Orbs* was finally published in 1543. The 'book that nobody read' featured a system of spheres in some ways just as complex and bizarre as the theories it was meant to supplant. But there was one crucial difference. The system centred on the Sun, and not the Earth. Copernicus may have made few observations, and his intentions may have been mystical rather than scientific, but it was his spirit of innovation that counted.

And so, Copernicus was to astronomy what Columbus was to navigation. Not its greatest exponent, but one who had the courage to confront his convictions, and see them through. It is the spirit at the very heart of the Renaissance, and one that marks such a revolutionary break with the aimless Middle Ages.

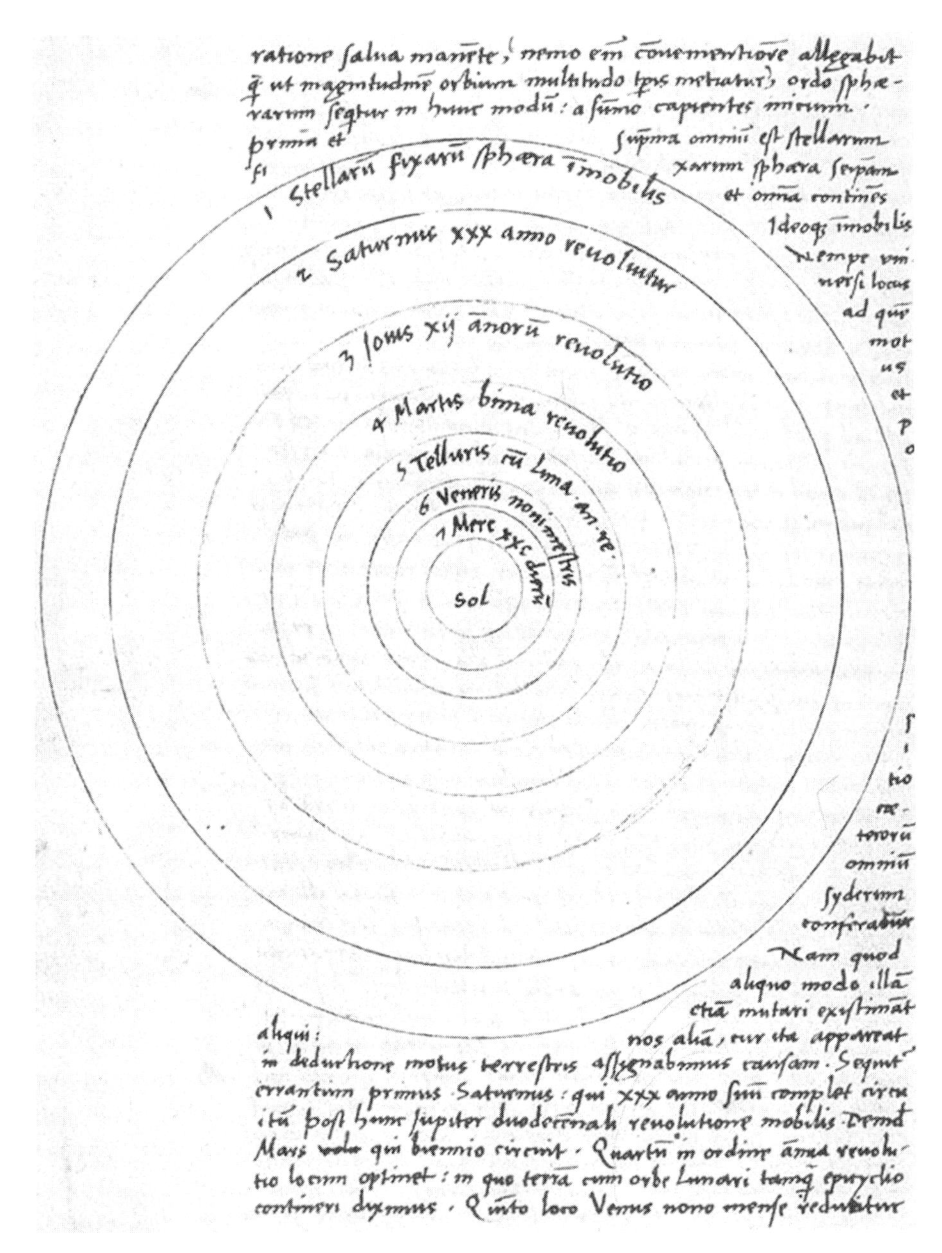

Figure 6 The Sun-centred system of Nicholas Copernicus, from *De Revolutionibus Orbium Coelestium*, Nuremberg, 1543. This rather prosaic woodcut set out the revolutionary doctrine of a moving Earth before Renaissance science and culture. A call to arms that others would later follow, and a philosophical stance that would hasten man's dethronement from the centre of the cosmos.

Now, for the first time in written history, the Sun and planets are set out in their true heavenly order. And yet the primary motives for Copernicus' revolutionary change were not scientific, but philosophic and aesthetic.[18] As Copernicus says of the simplicity of his heliocentric system,

> I think it easier to believe this than to confuse the issue by assuming a vast number of Spheres, which those who keep Earth at the centre must do. We thus rather follow Nature, who producing nothing vain or superfluous, often prefers to endow one cause with many effects.[19]

Once he has described each planetary orb in order, Copernicus then wraps up,

> In the middle of all sits the Sun enthroned. In this most beautiful temple could we place this luminary in any better position from which he can illuminate the whole at once? He is rightly called the Lamp, the Mind, the Ruler of the Universe; Hermes Trismegistus names him the visible God, Sophocles' Electra calls him the All-seeing. So the Sun sits upon a royal throne ruling his children the planets which circle round him. The Earth has the Moon at her service. As Aristotle says, in his de Animalibus, the Moon has the closest relationship with the Earth. Meanwhile the Earth conceives by the Sun, and becomes pregnant with annual rebirth.[20]

AD 1543: the date that marks one of the great turning points in the history of science.[21] [22] [23] This infamous book, *De Revolutionibus Orbium Coelestium*,[24] has the Sun at the centre of its planetary system, a scandalous and heretical demotion of the position of the Earth to that of mere planet.

The alien had arrived. For Copernicus set in train a revolution in astrobiology. A new physics was born, and a new mantra: if the Earth is a planet, then the planets may be Earths; if the Earth is not central, then neither is humanity.[25]

Crisis of the imagination

It was a slow burn start, this Copernican dawn.

At first the world held its breath. The notion of the new *Solar System* took time to catch, and the revolution was held in suspended animation. A vanguard of astronomers saw in the new system an elegance and simplicity for refining calculations. The German mathematician Erasmus Reinhold, for example, one of the leading astronomy teachers of his generation, produced *Prussian Tables* based on the Copernican model as early as 1551.

Beyond this inner circle of scholars, however, few people believed the new system was actually true. To begin with, the whole affair was an affront to common sense. And it seemed unthinkable that the entire Earth could revolve and rotate without warping the fall of shot, or whipping up a mighty wind – objections Galileo would presently remove.

Soon the notion spread. The very idea of an open cosmos, one in which the Earth was a mere atom, was contagious. It was bound to corrupt the old and decrepit cosmos of crystalline spheres, divinely made and maintained in motion. With this concept of a fresh and open universe grew another germ of an idea: the new learning had led to new worlds on Earth; might there not also be new worlds in the sky?[26]

And yet, Copernican cosmology cut two ways. It gave fresh impetus to the role of the imagination, even as it implied its moment of crisis. On the one hand, Copernicus' reorganisation of the cosmos, in which the Earth had been demoted to a mere planet, implied the possible existence of other Earths in a vast and boundless universe. It pulled the cosmos down to Earth, so to speak, bringing it into the realm of the imagination. And on the other hand, a doubt arose that, in a new universe not centred on the Earth, the human imagination may be ill equipped to allow access to these other realities. As Galileo put it, in his *Dialogue*,

> Thus, and more so, might it happen that in the Moon, separated from us by so much greater an interval and made of materials perhaps much different from those on Earth, substances exist and actions occur which are not merely remote from but completely beyond all our imaginings, lacking any resemblance to ours [our world] and therefore being entirely unthinkable. For that which we imagine must be either something already seen or a composite of things and parts of things seen at different times; such are sphinxes, sirens, chimeras, centaurs, etc.[27]

Doubt was everywhere.

The new system of Copernicus, and the theories it inspired regarding the boundless new universe with its plurality of worlds, gravely disturbed Western philosophy and religion. It raised doubts about articles of Christian faith, such as the doctrine of salvation and the conviction of divine power over all earthly matters. And it called into question the nature of Creation, and its relation to the Creator. In short, Copernicanism raised doubts that were of fundamental importance to human identity.

It was a crisis of idealism. In the age of faith, the imagination was thought to involve only relations between humans, their worldly objects, and the divine. The progress of materialist thought, however, and in particular the new cosmology, had profound implications for this traditional view of the imagination, and its creative capacity. Plato and Aristotle had prevailed because the human message of Jesus could not possibly have continued to serve the purpose of the Church once it sought cultural domination and power. Plato was resurrected. His philosophy was there in the message of St Paul, and as the gospel of St John shows, with its cult of the divine word, the *logos*, that Platonism was key from the very beginning.

Science and the imagination were in stasis. The rise of Christianity meant that, from the fourth century on, intellectual life was spoken through Christian thought. The Neo-Platonist thinker St Augustine had brokered a settlement between philosophy and faith, and was scornful of pagan science,

> When, then, the question is asked what we are to believe in regard to religion, it is not necessary to probe into the nature of things, as was done by those whom the Greeks call *physici*; nor need we be in alarm lest the Christian should be ignorant of the force and number of the elements: the motion, and order; and eclipses of the heavenly bodies; the form of the heavens; the species and the natures of animals, plants, stones, fountains, rivers, mountains; about chronology and distances; the signs of coming storms; and a thousand other things which those philosophers either have found out, or think they have found out . . . It is enough for the Christian to believe that the only cause of all created things, whether heavenly or earthly, whether visible or invisible, is the goodness of the Creator, the one true God; and that nothing exists but Himself that does not derive its existence from Him.[28]

Knowledge was confined to churchmen, and during the Middle Ages the history of the imagination over the lands of the disappearing Roman Empire was the history of Christian dogma. The Church fathers had set about a mission impossible. Their aim: to assimilate the more innocuous and idealist elements of the ancient wisdom into Christianity.

By stealth, much of classical philosophy had already found its way in. But the Old Testament and Greek culture were queer bedfellows. And as the Neo-Platonist Christians tried to crowbar in some safer aspects of philosophy, controversy was inevitable. The absurdity of fusing classical thinking with the mythical Testaments, Old and New, was bound to produce fatalities. Scripture is at best self-contradictory, and was never meant as a handbook of nature. The sheer nonsense of trying to merge the immiscible, in the form of natural philosophy and scripture, was lethal to any lucid grasp of the new universe, real or imagined. Faith and reason are irreconcilable. They cannot be squared without allegorising the one or mangling the other.

So it was that Aristotle's cosmology was writ large by Dante.

This finite geocentric cosmos endorsed by the Church was a black amalgam of Platonic philosophy, Aristotle's cosmology, and Christian dogma. Its most obvious feature was the opposition of a terrestrial realm and a spiritual one in the heavens. The contrast was stark. The distinction between Heaven and Earth was marked by respective oppositions such as timeless against transitory, immutable versus changing, and perfect against degenerate. The celestial realm, containing as it did the planets and the fixed stars, was governed by different laws, and shared in the nature of the divine: perfect, timeless, and immutable. As the heavens were made of an essence so superior and radically pure, they defied any comparison with earthly matter.

And so the belief was that the heavens, as a physical world, were just as unfathomable for the human mind as was the transcendental realm of God himself. The relation between terrestrial nature and divine truth was mystical. Knowledge of the godly realm was only really achieved in the hereafter. During their life on Earth, humans could only gain a limited knowledge of the nature of the cosmos to the extent that their flawed faculties allowed. They could certainly not take their perceptions at face value. Heaven forbid.

Remember, this was a world of limited horizons, and limited imagination. One might object, when considering the idea that space was beyond the imagination at this time. It seems rather unbelievable to us that reasonable, educated men and women would think in this nonsensical way. And yet this period is vividly rich with examples of the 'common sense' worldview of the time. To take but a select few: in 1458, a pig was hanged for murder in Burgundy; in 1602, the French judge Henri Boguet declared an apple possessed by demons; and a few years later, Italian Jesuits tried to calculate the physical dimensions of hell.[29] All witness to an imagination involving only worldly matters and the divine.

In such a culture, the heavens seemed beyond imagining. And there was little point in trying to describe them, at least in a physical sense. In essence, the cosmos either remained unknowable, or was misleading. So, rather than understanding the physical essence of the matter in the universe, it was better to know the spiritual significance of the cosmos, in all its oneness.

The world turned upside down

The one man who was to shine with the latent power of the imagination was himself a Copernican. And it was a heresy for which Giordano Bruno was burnt at the stake.

Bruno was an Italian philosopher, born near Naples just five years after the publication of Copernicus' great book. He was soon distinguished for his outstanding ability. During his time in Naples, Bruno became known for his skill with the art of memory, travelling to Rome to demonstrate his mnemonic system before Pope Pius V. But he was also possessed of a penetrating imagination, and a fiery temperament. Bruno dazzled merchants and scholars alike. Trouble was never very far away, however, for Bruno's acerbic tongue meant he was always on the move. He fell out with the monastic order to which he belonged, and his wanderings began in earnest.

The detail of Bruno's travels reads like a magical mystery tour of Renaissance culture. Tutored privately at an Augustinian monastery in Naples, at the age of seventeen he entered the Dominican Order. His taste for free thinking and forbidden books soon caused him difficulties. When his annotated copy of the banned writings of Erasmus was found hidden in the convent privy, Bruno fled, shedding his monastic habit.

Bruno began to shine. His travels through the Genoese port of Noli, then Savona, Turin, Venice, Padua, and Geneva, finally ended with a doctorate in theology and a lectureship in philosophy at Lyon. His meteoric rise to fame continued in Paris, where his incredible feats of memory were attributed to Bruno's magical powers. Indeed, his published work on mnemonics, *De Umbris Idearum* (*On the Shadows of Ideas*, 1582), was dedicated to King Henry III of France. Given that sixteenth-century dedications were approved in advance, it is clear that Bruno had begun to move in powerful circles.

In 1583, Bruno travelled to England. He had done so on the counsel of Henry III, and as a guest of the French ambassador, Michel de Castelnau. While in England he met with one of the Elizabethan age's most prominent figures, the English poet Philip Sidney, to whom Bruno dedicated two books. Bruno also became friendly with members of the Hermetic circle. The circle was focused around John Dee, sometime Welsh physician to Queen Elizabeth I, and another mystic who dangerously straddled the worlds of science and magic, just as they were becoming distinguishable.

Bruno became a Copernican. Whilst lecturing at Oxford, he failed in his attempt to secure tenure at the university, but was successful in courting controversy with potentially powerful priests. Bruno's spat involved John Underhill, Rector of Lincoln College (and from 1589 Bishop of Oxford), and George Abbot, who was soon to become Archbishop of Canterbury. It was Abbot who first derided Bruno for holding, 'the opinion of Copernicus that the earth did go round, and the heavens did stand still; whereas in truth it was his own head which rather did run round, and his brains did not stand still'.[30]

Even so, Bruno's time in England was fruitful, resulting in some of his most important works. Among these his *Italian Dialogues* is significant, for the book includes cosmological tracts such as *La Cena de le Ceneri* (*The Ash Wednesday Supper*, 1584), *De la Causa, Principio et Uno* (*On Cause, Principle and Unity*, 1584), and *De l'Infinito Universo et Mondi* (*On the Infinite Universe and Worlds*, 1584).

Bruno became a committed pluralist.

His *On the Infinite Universe and Worlds* was a form of cosmic pantheism, believing God and universe to be one and the same thing. Bruno's imagination populated planets and stars, attributing them souls, and even doing the same for the cosmos as a whole. In Bruno's words, 'Innumerable suns exist; innumerable earths revolve around these suns in a manner similar to the way the seven planets revolve around our sun. Living beings inhabit these worlds.'[31] As the American political leader Robert Green Ingersoll was to say of Bruno,

> The First Great Star, Herald of the Dawn, was Bruno ... He was a pantheist, that is to say, an atheist. He was a lover of Nature, a reaction from the asceticism of the church. He was tired of the gloom of the monastery. He loved the fields, the woods, the streams. He said to his brother-priests: Come out of your cells, out of your dungeons: come into the air and light. Throw away your beads and your crosses. Gather flowers; mingle with your fellow-men; have wives and children; scatter the seeds of joy; throw away the thorns and nettles of your creeds; enjoy the perpetual miracle of life.[32]

With Giordano Bruno, Copernicanism had truly taken flight. For Bruno's cosmology was far more radical than that of Copernicus: a rotating and revolving Earth was responsible for the illusion of the night sky; the cosmos was infinite in extent; and Bruno saw no reason to believe that the stars were equidistant from a single centre of the new universe.

Bruno abandoned Aristotle. The Earth was a mere heavenly body, as was the Sun. God had no more business with the Earth than he did any other far-flung planet. According to Bruno, God was as present in the heavens as he was on Earth. God was an Immanence, subsuming the great multiplicity of existence, instead of the rather miserably remote heavenly deity championed by the Church.

The cosmology of Bruno was Atomist. His was an infinite universe, in a constant state of flux. Space and time were endless, and identical physical laws governed every last corner of the cosmos, a cosmos made everywhere of the four elements, rather than having the stars composed of a different quintessence. In Bruno's vision, there was no room for stability and permanence, for Christian notions of Divine Creation and a Last Judgement. Rather, the Sun was simply a star, not at all the high-flown mystic body of Copernicus' vision. And all the stars were suns. The cosmos was an infinite number of solar systems, the basic building block of the entire universe. Matter itself was made of discrete atoms, separated by vast reaches of ether, as he believed empty space could not exist.

Bruno's ideas were held up for ridicule. On returning to Padua, he taught briefly, and then tried unsuccessfully for the chair of mathematics, a post that was to be assigned instead to Galileo a year later. After his last years of wandering, the Inquisition finally summoned Bruno to Rome. He was imprisoned for seven years during his prolonged trial, lastly in the Tower of Nona, and the ill-famed dungeon's terrible lightless cells, one of which was known as 'the pit'. Some key papers about the infamous trial are lost, but a stunning find was made in 1940, when the summary of the trial's proceedings was unearthed.[33]

The Inquisition nailed Bruno. The charges naturally included heresy, blasphemy, and immoral conduct. But, most significantly for our story, especially in view of what was to come later with Galileo, Bruno was charged on the basis of his doctrines of philosophy and cosmology. The charges included holding the existence of a plurality of worlds and their eternity to be true; holding opinions contrary to the Catholic faith and speaking against it and its ministers; holding flawed opinions about the Trinity; holding invalid opinions about Christ; denying the Virginity of Mary; and dealing in magic and divination. Confronted with this hopeless state of affairs, Bruno bravely set out his defence. Remarkably, he was ready to bow to the Church on its dogmatic teachings, but held firm on his philosophy. Most of all, his Copernican belief in the plurality of worlds he held most dear, even though he was told to abandon it.

At last, on 10 February 1600, Bruno was led out to the Church of Santa Maria, and sentenced to death. As the Holy Church was curiously to phrase it, he was to be punished 'as mercifully as possible, and without effusion of blood'. In defiant reply, Bruno's famous response was, 'You are more afraid to pronounce my sentence than I to receive it'.[34] He was allowed a week's grace for recantation, but to no avail. On 17 February 1600, Bruno was burnt to death on the Field of Flowers. To the end he was brave and rebellious. Legend has it he scornfully pushed aside the crucifix they gave him to kiss. As one of his adversaries admitted, Bruno died calm and heroic, defiant of wrong.

Freethinker and secularist George William Foote found great inspiration in Bruno's example,

> Such heroism stirs the blood more than the sound of a trumpet. Bruno stood at the stake in solitary and awful grandeur. There was not a friendly face in the vast crowd around him. It was one man against the world. Surely the knight of Liberty, the champion of Freethought, who lived such a life and died such a death, without hope of reward on earth or in heaven, sustained only by his indomitable manhood, is worthy to be accounted the supreme martyr of all time. He towers above the less

> disinterested martyrs of Faith like a colossus; the proudest of them might walk under him without bending.[35]

Despite the intervening centuries, controversy still rages over the reasons for Bruno's death, and over his contribution to the scientific revolution. At first, his ideas continued to be held up for ridicule. Prolific seventeenth-century writer Margaret Cavendish, Duchess of Newcastle-upon-Tyne, penned a run of poems against 'atoms' and 'infinite worlds' in her *Poems and Fancies* in 1664. But Bruno's potential began to shine through with implications for Newtonian cosmology, an infinite universe lacking in structure, and a crucial cross-point between the old and the new.

Debatably, far more convincingly than Copernicus, it is the cosmology of Bruno that genuinely signals the paradigm shift of the old universe, and into the new universe. Aristotle's cosy geocentric cosmos was about us. The new universe was decentralised, inhuman, and alien. Through Bruno's words, we can more clearly see the cultural shock created by the discovery of our new and marginal position in a cosmos fundamentally inhospitable to man.[36]

Bruno's universe is remarkably fresh. It is infinite, homogeneous, and isotropic, or uniform in all directions. Planetary systems are distributed evenly throughout the cosmos. It is even possible to see in Bruno's idea of multiple worlds, and the infinite possibilities of the indivisible One, a forerunner of the many-worlds interpretation of quantum mechanics.

As for Bruno's trial, there are clear parallels with the more famous case of Galileo, which was to follow a few decades later. The 'Hammer of the Heretics', inquisitor Cardinal Bellarmine, oversaw Bruno's trial. It was Bellarmine who demanded a full recantation, which Bruno eventually refused. And Bellarmine also played a crucial role in the trial of Galileo.

Many see Bruno, and not Galileo, as the true and original martyr of science, identifying his Copernican pluralism as an important factor in the outcome of his trial. As John J. Kessler writes, 'He is one martyr whose name should lead all the rest. He was not a mere religious sectarian who was caught up in the psychology of some mob hysteria. He was a sensitive, imaginative poet, fired with the enthusiasm of a larger vision of a larger Universe... and he fell into the error of heretical belief.'[37]

There is a view among some that Bruno's death is somehow unconnected with his pluralist cosmology. To bolster this belief, they point out that in 1600 the Catholic Church had no formal policy on the Copernican system, and that it was certainly not heresy.[38] [39] Copernicanism, they propose, was not specifically proscribed as heretical until well after Bruno's death, when Bellarmine himself placed Copernicus' *De Revolutionibus*, among others, on the 'Index of Forbidden Books'.

And yet there is far more in heaven and hell than the documentary evidence of a wilful bureaucracy. The nascent ideas of Atomism, Copernicanism, and pluralism play a crucial part of the dramatic trials of both Bruno and Galileo. Moreover, the secret archives of the Vatican tell a different tale. In the detail of Bruno's trial in Rome, the documents state,

> In the same rooms where Giordano Bruno was questioned, for the same important reasons of the relationship between science and faith, at the dawning of the new astronomy and at the decline of Aristotle's philosophy, sixteen years later, Cardinal Bellarmino, who then contested Bruno's heretical theses, summoned Galileo Galilei, who also faced a famous inquisitorial trial, which, luckily for him, ended with a simple abjuration.[40]

It was for his belief in aliens, along with other alleged heresies, that the Roman Catholic Church burned Giordano Bruno at the stake. He certainly suffered a cruel and unnecessary death. Some say it was a martyr's death. Others say that Bruno, and not Galileo, has become the Church's most difficult cause célèbre:

> The murder of this man will never be completely and perfectly avenged until from Rome shall be swept every vestige of priest and pope, until over the shapeless ruin of St. Peter's, the crumbled Vatican and the fallen cross, shall rise a monument to Bruno, the thinker, philosopher, philanthropist, atheist, martyr.[41]

It is time to consider the telescope.

The far-seer

Giordano Bruno stands at a cusp.

He stands at the boundary between pre- and post-Copernican worlds. Bruno's is a rather magical art. Like John Dee and Paracelsus, he dabbled in Hermetic magic. For Bruno, with the help of memory, imagination was 'man's highest power, by means of which he can grasp the intelligible world beyond appearances through laying hold of significant images', stored in his memory.[42] So in this respect Bruno believed that man could only hope to gain *spiritual* access to celestial knowledge.

In this way Bruno's belief was that the universe was essentially animistic, and that the motion of matter was due to the movement of the souls and spirits of which matter was made. His approach also seems to have disregarded mathematics as a means to understanding. It is perhaps the most dramatic way in which Bruno's cosmology differs from today's state-of-the-universe picture.

And yet, Bruno was also a Copernican. The new theory demoted the Earth as the centre of the universe, giving it the far more lowly status of planet. Equally significant, it introduced the possibility of a material similarity between the Earth and the other planets, which all revolve around the Sun. Indeed, Bruno was quick to realise the bigger picture too. The heavens were now no longer imbued with the profound religious sense of 'otherness'. They became thinkable as a physical world, just like Earth. And Earth itself no longer served as the point of reference for the whole cosmos. The universe became decentralised.

The alien planet was born. With our world bursting open, the search for the new was directed outwards, and turned into a quest for the discovery of new worlds. The opposition between terrestrial and celestial realms disappeared. In its place evolved not just the idea of cosmic similarity, but a new potential for *difference* that emerged with the possibility of alternative worlds.[43] But how would such *difference* be imagined?

The telescope was to prove crucial for Copernicanism. No scientist was needed to invent it. The device was long overdue, as the means of making the so-called spyglass had probably been present for at least three hundred years. Only with the sheer volume of lens manufacture through the increased wealth of the sixteenth century was the discovery brought about by chance. So, rather than any further refinement to theory, it was the inception of the *far-seer*, which was to bring the heavens down to Earth, and the old cosmology of Aristotle to its knees.

It was in March 1610 that Galileo's revolutionary discoveries with the spyglass were hurled like an incendiary device into a rather sleepy and dull academic world. Inside his brief but magnificent flyer, the momentous and now legendary book *Sidereus Nuncius* (*The Starry Messenger*), Galileo laid out the most incredible evidence for a new Copernican cosmos.

The book was a revolution in the communication of science. It was radical, not only in content but also in terms of its brief twenty-four pages. Remarkably, *The Starry Messenger* was Galileo's first scientific publication. It boasted of discoveries, 'which no mortal had seen before',[44] and was penned in a new, tersely written way, which few scholars had used before. The initial print run was 550 copies, sent out to key figures in the universities of Italy, and also to booksellers in every major city. The result was culture shock. The sophisticated Imperial Ambassador in Venice said the book was 'a dry discourse or an inflated boast, devoid of all philosophy'.[45] Nonetheless, the book created the dramatic effect for which it was designed, arousing immediate and passionate controversy.

Galileo's *The Starry Messenger* begins with his observations of the Moon:

> The surface of the Moon is not perfectly smooth, free from inequalities and exactly spherical, as a large school of philosophers considers with

> regard to the Moon and the other heavenly bodies, but that, on the contrary, it is full of irregularities, uneven, full of hollows and protuberances, just like the surface of the Earth itself, which is varied everywhere by lofty mountains and deep valleys.[46]

The quote makes Galileo's intentions clear. Aristotle had been the last notable ancient cosmologist, his vision of a lifeless cosmos casting a long, dark shadow over astronomy for millennia. But Galileo shows how Aristotle's system does not stand up to modern scrutiny.

Consider the evidence, Galileo says. In Aristotle's two-tier cosmos, only the Earth was meant to be subject to the horrors of change, death, and decay. Beyond the Earth, the celestial sphere was immutable and perfect. And yet, through the spyglass, the picture is dramatically different. The Moon is pock-marked and bumpy, far from crystalline and perfect. As for Aristotle's claims that the Earth is unique, Galileo shows that the Moon has Earth-like mountains, valleys, and hollows.

What life lies beyond?

There was another decisive way in which these discoveries were revolutionary: you could see for yourself. In the age of faith, the enemy was monopoly. For centuries, knowledge had been shut up in the Latin Bible, which only black-coated ministers were to interpret. Now, if you failed to believe what Galileo said of the lunar surface, anyone with enough ready cash to do so could pop out to the local spectacle-maker, pick up a spyglass, and see the surface of the Moon for himself.

The cosmos opened up before all those who were brave enough to look. In *The Starry Messenger*, Galileo had also turned his gadget to gaze at the stars, describing how the spyglass prised open the heavens for deeper exploration, 'other stars, in myriads, which have never been seen before, and which surpass the old, previously known stars in number more than ten times'.[47] To the nine stars in the belt and sword of Orion, Galileo added eighty more; to the seven stars in the constellation of Pleiades, the seven sisters, he made thirty-six; and as for the belt of the Milky Way, the telescope resolved the Galaxy into 'a mass of innumerable stars planted together in clusters'.[48]

A question in the nerves was lit.

If the magnificent splendour of Aristotle's God-given cosmos is made especially for man's delight, why is it only through a machine, such as the spyglass, that man can justly savour its intricate parts, and begin to know its true nature? Furthermore, the telescope had revealed a new and vast cosmos that stretched beyond the reach of human vision. It robbed the old universe of its unity.

The old cosmos was transformed. Consider the Moon: with its prominent position in the night sky, it was the focus of exploration with the new far-seer. And it was unique among the heavenly bodies in that its image through the spyglass unveiled features that made it easy to describe by terrestrial analogy. The Moon was evidence that there was material similarity between the Earth and the rest of the universe. The Moon became the focus for the alien fiction of the age. These early off-world narratives were explorations of alien life, space voyages such as Johannes Kepler's *Somnium* (*The Dream*) (1634) and Francis Godwin's *The Man in the Moone* (1638).

The dark side of Moon

Kepler's story *Somnium* is an extended voyage of discovery, from the oceans of Earth to the dark seas of the Moon. Galileo's *Starry Messenger* had inspired in Kepler early and uncannily prophetic ideas of alien life. In his first hasty response, *Conversation with the Starry Messenger*, Kepler had hinted at his fascinating insight into Copernicanism, alien life, and space travel:

> There will certainly be no lack of human pioneers when we have mastered the art of flight. Who would have thought that navigation across the vast ocean is less dangerous and quieter than in the narrow, threatening gulfs of the Adriatic, or the Baltic, or the British straits? Let us create vessels and sails adjusted to the heavenly ether, and there will be plenty of people unafraid of the empty wastes. In the meantime, we shall prepare, for the brave sky-travellers, maps of the celestial bodies – I shall do it for the Moon, you Galileo, for Jupiter.[49]

And yet Kepler already had his own startling story to tell. The year before Galileo's earth-shattering discoveries with the spyglass, Kepler had published in draft manuscript what many regard as the first ever work of science fiction. *Somnium* was a space voyage of discovery in the new physics that invented the alien, and spoke of a cosmos that was soon to be unveiled by the telescope. As one commentator put it, 'Kepler's *Somnium* is presented as the product of the occupations of the day, since it melds into a single narrative the "residue" of the reading of a magician's story and observations through a telescope'.[50]

Copernicus had shifted the hub of the universe to the Sun. Kepler's aim was to explore the new and alien panorama from the alternative standpoint of the Moon. He wanted to show that a living being on the Moon would develop an astronomy that takes the Moon as the stable centre of the cosmos, around which all else moves, orbiting Earth included. Indeed, they would not only be likely to conceive of the cosmos as lunocentric and lunostatic, they would, besides, witness a spectacle that to humans on Earth was so difficult to conceive: the Earth spinning around, once every twenty-four hours.

In this way, the Copernican system could be compared and explored. The Moon becomes a mirror of faulty terrestrial science. The lunocentrism and lunostatism of lunar creatures matches the geocentrism and geostatism of narrow-minded Earth dwellers. The Moon orbiting the Earth becomes a fictional device of the Earth orbiting the Sun. And novel observations of the heavens, and the Earth itself, are made viable.

But Kepler had to be careful. The world was watching. He chose a dream story for his Copernican tale, to subvert the wrath of the Aristotelians by concealing his science in the guise of myth. Kepler got new ideas by reading other fictions, though his plan to make a dream narrative was sometimes governed by pure chance. Kepler relates how he read Lucian's *True Story* to teach himself Greek, but was then gripped by this account of a voyage to the Moon, 'He [Lucian], too, sails out past the Pillars of Hercules into the ocean and, carried aloft with his ship by whirlwinds, is transported to the Moon. These were my first traces of a trip to the Moon, which was my aspiration in later times.'[51]

Though the tale was in draft manuscript form in 1609, Kepler had actually been working on the text since 1593. *Somnium* is a truly radical work. Its theme is the new cosmos of Copernicus. There is rarely an extract in Kepler's bountiful output that is not alive with imagination. So it is with *Somnium*. As a fictional travelogue it follows on from Lucian's *A True Story*. In all other ways it is markedly different.

The journey to the Moon is inventive genius. Even Kepler's use of myth is cunning and creative. On his space voyage, the 'spirit of the Moon' Daemon carries Kepler's hero, Duracotus, to a lunar landscape. But first they use a piece of Pythagorean daemonology to get there. The Daemon explains that he and his kindred spirits can only embark on the journey during a lunar or solar eclipse, when the cone of the shadow, between the Moon and the Earth, serves them as a ladder.

On arriving, he gazes down upon the Earth, noting its geography, its movement through space, and exploring the surface of the Moon, and its alien inhabitants. Kepler's mastery of Copernicus is evident throughout *Somnium*. Although Duracotus initially journeys to the Moon by mystic means, physical law soon kicks in, and fantasy makes way for science:

> The journey ... becomes easier because ... the body ... escapes the magnetic force of the earth and enters that of the moon, so that the latter gets the upper hand ... At this point we set the travellers free and leave them to their own devices ...[52]

In his 1609 opus *Astronomia Nova (A New Astronomy)*, Kepler had hinted at the concept of gravity. In *Somnium*, he makes his ideas clear. With astonishing prescience, he also proposes the existence of micro-gravity zones: 'for, as magnetic forces of the earth and moon both attract the body and hold it suspended, the effect is as if neither of them were attracting it'. Kepler's insight does not stop there. He assumes there are spring tides on the Moon, due to the joint attraction of Sun and Earth. Duracotus' journey takes the shortest route to the Moon; not the straight line used by Lucian and Plutarch, but a trajectory from Earth to a point in space where the Moon and the lunar voyagers would arrive simultaneously.[53]

Presently, the speed of flight carries the travellers 'along almost entirely by our will alone, so that finally the bodily mass proceeds toward its destination of its own accord'. As John Lear pointed out in his introduction to the first English translation of *Somnium*, Kepler had pioneered the concept of 'inertia' and taken it into outer space. Kepler's is the inertia of rest, to which Newton was to add the inertia of motion. Kepler also understood the dangers of lunar flight. He believed it was scientifically possible to reach the Moon, again distinguishing his work from the writers of the past.[54]

The killing Moon

Once the journey is over, Kepler reveals his imagined Moon. The Daemon gives a detailed account of Levania, the Moon. It is a lunar landscape of Copernican precision. From sunrise to sunset, a Moon-day lasts for two weeks. The same goes for a lunar night, since the Moon spins on its axis once a month. As a lunar orbit also takes a month, the Moon always shows the same face to the Earth. The alien creatures of the Moon call Earth 'Volva' (as in *revolvere*, to turn). The Earth-facing half of their home is known as the Subvolvan, the far side as the Privolvan – both halves have a year of twelve days and nights.

Kepler's Moon is a truly alien world.

His nightmare lunar vision begins with the deadly differences in temperature – blazing days, freezing nights – that plague the landscape. The starry canopy above the Moon is equally strange: across a bible-black sky, the stars, Sun, and planets scuttle relentlessly to and fro, due to the Moon's orbit about the Earth. Indeed, this bewitching 'lunatic' astronomy of Kepler's is yet to meet its match in fiction.

The Daemon describes life on Levania in detail. Topographically, the lunar world is much like Earth, though everything is on a grander scale: their mountains soar to incredible heights, their valleys deeper than any earthly canyon. Despite Galileo's holding an opposite view, Kepler imagined water on the Moon, and hence the existence of life.

The extraterrestrials that stalk Kepler's lunar world are not men and women, but creatures fit to survive the alien haunt. Two centuries before Darwin, Kepler had already understood the link between lifeform and environment,[55] 'nothing on Earth is so fierce that God did not instill resistance to it in a particular species of animal: in lions, to hunger and the African heat; in camels, to thirst and the vast deserts of Palmyrene Syria; in bears, to the cold of the far north, etc.'[56] And while studying at university, Kepler had written:

> If there are living creatures on the Moon (a matter which I took pleasure in speculating after the manner of Pythagoras and Plutarch in a disputation written in 1593), it is to be assumed that they should be adapted to the character of their particular country.[57]

Indeed, Kepler's approach is distinctly Darwinian. The *Somnium* narrative declares, 'In plants, the rind; in animals, the skin, or whatever replaces it, takes up the major portion of the bodily mass; it is spongy and porous. If anything is exposed during the day, it becomes hard on top and scorched; when evening comes, its husk drops off.'[58] Kepler explains, in the appendix, the inspirational source that fed his imagination in the creation of lunar life: 'The precedents were our vegetables' and fruits' rinds, varying with nature's varying providence; the precedents were oysters' and turtles' shells, shaped like shields; the precedents were the calluses on the feet, the hoofs and the soles of animals.'[59]

Kepler's killing Moon is hardest on the Privolvans.

They suffer relentlessly long nights, 'bristling with ice and snow under the raging, icy winds', with no reprieve gifted by Earthlight, as they are not able to bask in it. The Privolvan 'day' is little better: for two weeks at a time, the scorching Sun bakes the lunar air, so that it is 'fifteen times hotter than our Africa'. The atrocity of the Privolvan climate makes a settled form of existence impossible, and so they boast 'no fixed and safe habitations; they traverse in hordes, in a single day, the whole of their world, following the receding waters either on legs . . . or on wings, or in ships.'

Privolvan territory is thus compared to earthly wastes, forests, and deserts. And the physiology of Privolvan creatures is adapted to their conditions: some have legs, 'which far surpass those of our camels',[60] some resort to wings, while others, helping nature with art, follow the receding water in boats. 'Most of them are divers; all of them, since they live naturally, draw their breath very slowly; hence under water they stay down on the bottom.'[61] Indeed, the Daemon's original flight from the Sun was into the 'caves and dark places' of the Moon, which parallels the frantic struggle for life of the lunar creatures, who spend half their life underground, trying to escape the searing heat.

This prescient idea of the adaptation of life to environment is a central feature in Kepler's imaginative creation of Privolvan forms of life:

> Since I had deprived the Privolvans of water, and I was compelled to leave them immense alterations of heat and cold coming directly on each other's heels, it occurred to me that those regions could not be inhabited, at least not out in the open. Hence it was convenient that water flowed in at fixed times of the day [Kepler here refers to lunar tides he has bestowed]. When it receded, I had the living creatures accompany it. To enable them to do so quickly, I gave some of them long legs and others the ability to swim and endure the water, with the proviso that they would not degenerate into fishes. None of this will be unbelievable to anybody who has read about Cola, the Sicilian man-fish.[62 63]

The Privolvans turn into Sun-worshippers, as Kepler offers evolutionary options. In addition to making the Privolvans fit enough to seek cover from the searing Sun, he suggests they may also be fully adapted to the lunar climate. For this reason, he suddenly declares in the narrative of *Somnium*, 'In general, the serpentine nature is predominant. For in wonderful manner they expose themselves to the sun at noon as if for pleasure.'[64]

Meanwhile, in the Subvolvan hemisphere, the nights are softened by the heat and light of the Earth, which stands still in the sky, 'as if nailed on'.[65] The Subvolvan territory is comparable to Earth's cantons, towns, and gardens. Like Earthlings, the Subvolvans see the spectacle of the waxing and waning of the Earth's surface, fifteen times the size of our Moon. And they gain respite from the constant cover of clouds, whose rain gives relief from the heat. The mountains of the Moon climb higher than those of Earth. Kepler's creatures attain a monstrous size. They feed only at night, for feeding after sunrise often leads to death. The hide of these massive serpents is permeable and, if exposed to the full force of the Sun, becomes seared and brittle. Such is life for this gigantic race of short-lived beasts. They bask for a fleeting instant in the rising or setting Sun, then creep into the impenetrable lunar darkness.[66]

Somnium is a turning point. It signals the end of the old era, and heralds loudly the dawning of the new science. Kepler's tale of extraterrestrial life had an estranging effect on its reader, revealing the world in a new light. *Somnium* is an imaginative tour de force. The book was the first space fiction of the age. And as we shall see, the alien voyages quickly evolved, from Kepler, through Francis Godwin, and on to Cyrano de Bergerac.

But *Somnium* is also a beacon. It is a fictional exercise in Copernicanism, as Kepler's tale does not only imagine life on our lunar world, reflecting the Moon's cosmic position and climate. It also assumes that lunar life will spawn out of the principle of environmental adaptation, as does life on Earth. The terrestrial and lunar worlds are analogous. Kepler's conclusion is Atomist in its pluralism: any part of the universe that is potentially comparable to Earth in nature is designed to accommodate life.

The influence of Kepler's *Somnium* was huge. It motivated much later tales of interplanetary journeys, such as those of H. G. Wells and Arthur C. Clarke. And it inspired a potent motif for exploring the insignificance of man. Kepler was a pioneer of the vision of space as the home of a plurality of inhabited worlds.[67] There is no better testimony to the power and imaginative sway of space fiction than this – despite the staggering odds against finding life beyond our planet, billions of dollars have been spent on sober scientific projects in search of alien intelligence. That search started with Kepler.

And the planets may be earths ...

Not everyone believed what they heard about Galileo and his discoveries with the 'optick tube'. Copernicus had been a quiet and humble man all his life. And no one who met the captivating Kepler could take exception to him. Galileo was quite another matter. Portraits show a red-haired and stocky man, with rough features and an arrogant stare. Galileo was politically radical, and had contempt for authority, which would later lead to terrible trouble.

Galileo's gift for vision promoted blindness in others. And his talent for antagonising peers led to scepticism in the small academic world of his own country. A case in point occurred in 1610, on 24 and 25 April. Evening parties were held to celebrate his recent discoveries with the far-seer. The great man was asked to demonstrate the Jupiter moons in the spyglass. Not one among the eminent guests was convinced of their existence. The crude nature of the mysterious gadget did not help. But many were blinded by bias – they refused to look down the tube.

And there was uproar. A public controversy followed, similar in some senses to the UFO farce, three hundred years later: allegations of optical illusions, haloes, and the unreliability of inexpert witness. Though Galileo and his telescope became the talk of the world's poets and philosophers, scholars in his own land were sceptical, or blatantly hostile. Kepler's was the only weighty voice raised in defence.

The revolution in science begins with the discovery of these 'alien' worlds, and the names of Kepler and Galileo have come to symbolise that revolution. They produced a map of the knowable, just as the unknown was at the point of becoming known. In speaking of sirens, centaurs, and chimeras, Galileo implied that Kepler's creatures may well be dwelling on the Moon. It was a critical new piece of evidence in the debate on the existence of extraterrestrial life.

The Moon became real. In many ways, Galileo's telescopic discovery of the lunar surface rendered that alien world a real object for the first time. But at that very moment, we also feel a sense of wonder, or *estrangement*, from this new reality. By estrangement we necessarily mean a sense of flawed knowledge, the result of coming to understand what is just within our mental horizons.[68]

This same sense of estrangement saturates the emerging space fiction. Little wonder that Galileo inspired Kepler to say the Jovian satellites (a word coined by Kepler himself) must be inhabited. As we shall see later, it was Kepler who motivated H. G. Wells to write, nearly 300 years later, 'But who shall dwell in these Worlds if they be inhabited? Are We, or They, Lords of the World? And how are all things made for Man?'[69]

The Moon was a touchstone. In thinking about the shadows on the lunar surface, Galileo had to assume they had similar causes to shadows on Earth, in order to understand the Moon's difference. And as eminent an astronomer as Kepler clearly needed to believe in extraterrestrials in order to make Galileo's discovery thinkable. Kepler understood that to know the Moon it was not enough to put the telescopic observations into words. The words themselves had to be transformed by a new sort of fiction.

So *Somnium* is revolutionary in the history of science. Throwing words at the Moon, as it were, has a dialectic effect – the words come back to us transformed. By imagining strange worlds, even those as commonplace as our Moon, we come to see science and life itself in a new perspective. Kepler's journey to the Moon was a very different piece of work compared with Lucian's *A True Story*. For one thing, it was written one and a half thousand years later. But the most striking difference was this: whereas Lucian's work is mostly fantasy, Kepler's *Somnium* was a conscious effort to address the new physics. And the crucial factor in the difference between the stories is the developed scientific consciousness evidenced by the newly invented telescope.

There was another key factor in the evolution from Lucian to Kepler: economy. The change in material conditions, such as the development of the printing press, had transformed the spread of knowledge. Publication rates rocketed. In England, for example, only eighty books were published each year in the 1540s. But by the 1640s, this figure had mushroomed to four thousand. The readership became far more diverse. Kepler was well aware of the rising use of print as an ideological battleground. By the 1620s the telling of sensationalist stories with a moral purpose became very popular, and Kepler's dream tale suited the traditional oral culture.

And so it began. This new breed of space fiction was launched along with the new physics. Both marked the most massive paradigm shift science has ever seen, as the old and cosy universe of Aristotle slowly morphed into the new universe of Copernicus; decentralised, infinite, and alien. The old universe had the stamp of humanity about it. Constellations bore the names of earthly myths and legends, and a magnificence that gave evidence of God's glory.

The new space fiction was a response to the new and inhuman universe. It dealt with the shock created by the discovery of our marginal position in a fundamentally inhospitable cosmos.[70] Earth was no longer at its centre, nor was it made of a unique material only to be found on terra firma. Nor was the Earth the only planet with bodies in orbit about it, as Galileo had discovered the Jovian moons.

The Man in the Moone

Kepler's *Somnium* was just the start. As space fiction matured, its attempts to make human sense of Copernicus' new universe became more sophisticated. Francis Godwin, the Bishop of Llandaff in Cardiff, Wales, became the first person to write a space travelogue in the English language. Godwin's lunar odyssey, *The Man in the Moone: or, A Discourse of a Voyage Thither, by Domingo Gonsales, the Speedy Messenger*, was another key work of early space fiction.

The Man in the Moone was published posthumously in 1638, but, like Kepler, the Bishop had started to write his Moon tale much earlier, in 1589. Godwin did not survive to see the book's outstanding success. Over the next century, *The Man in the Moone* went through two dozen editions, and was translated into many languages, including French, German, and Dutch. Indeed, Godwin's Moon tale was regarded as the seminal space voyage, with nineteenth-century writers such as Jules Verne and Edgar Allan Poe crediting him as a major inspiration.

Godwin's fiction also inspired the imagination of John Wilkins, one of the joint first secretaries of the Royal Society. Wilkins had published *The Discovery of a World in the Moone* (1638), which had proved a fashionable piece of seventeenth-century science communication. The book was Wilkins' contribution to the debate on the plurality of habitable worlds, covering a range of speculative topics, including the nature of the Moon and visible features on the lunar surface, Luna's likely inhabitants, and our possible means of space travel.

The third edition of Wilkins' book was revised to take account of the popularity of Godwin's work. The newly added fourteenth chapter ends,

> Having thus finished this discourse, I chanced upon a late fancy to this purpose under the fained name of Domingo Gonsales, written by a late reverend and learned Bishop; In which (besides sundry particulars wherein this later chapter did unwittingly agree with it) there is delivered a very pleasant and well contrived fancy concerning a voyage to this other world.[71]

One of the main mantras of Wilkins' book had been this simple but Copernican question, 'Now if our earth were one of the Planets (as it is according to them) then why may not another of the Planets be an earth?'[72] Wilkins' point was this. Though Copernican ideas may seem rather 'horrid', telescopic evidence seemed to support them. And if the ideas seem offensive at first, still we may get used to them, even comfortable with the concepts. Earthliness, he says, is a matter of degree, all solar and extra-solar planets lying along a spectrum from most Earth-like to least. So seeing the Moon as a type of planet is a triumph for the scientific imagination.

Godwin's *The Man in the Moone* is the first English book in history to portray alien contact. Wilkins, and Godwin himself, thought the main point of Godwin's book to be an account of the possibility of a space voyage to another Earth-like world. But the book became more renowned for the idea that alien contact was a real possibility; it was only a matter of time before such an encounter took place. Galileo's lunar discoveries with the spyglass are also clearly evident in the book, and the preface of Godwin's fiction credits Galileo as the discoverer of this new world in fact.

Francis Godwin was born in 1562, the son of Thomas Godwin, Bishop of Bath and Wells. Godwin was at Oxford when Giordano Bruno had delivered his sensational sermons on the question of life beyond the Earth. Perhaps Bruno inspired the idea in Godwin of an inhabited Moon, though it was through Galileo that the notion became widely known.

Once again, ships prove crucial in this age of the space voyage. Whilst at Llandaff, Godwin had profitable links with the thriving local sea trade. Merchants and sea captains out of the ports of Cardiff and Bristol dined with Godwin at the Bishop's Palace in Llandaff, a mere three miles from the docks at Cardiff. Admittedly, Godwin was entertaining to raise funds for the restoration of the cathedral. But he also listened to maritime tales of discovery and endeavour. Godwin's fiction is similar to the diaries kept by notable explorers of his time, such as Richard Hakluyt, a style that he may have copied.

The Man in the Moone is about the journey of an early space traveller, Domingo Gonsales. Señor Gonsales' voyage of discovery goes somewhat astray. His ultimate destination is China, but he takes a rather circuitous and accidental route via St Helena, and the Moon! For his initial voyage, Señor Gonsales has reared and trained forty wild geese ('gansas'), but on returning to Spain his ship is attacked and sunk by British privateers. Making good his escape, Gonsales harnesses himself to the birds, an exceptionally bizarre flying machine. The *gansas* carry him higher and higher, and a little too late he remembers they migrate each year to hibernate on the Moon.

His interplanetary flight is full of wonder. His experiences on the journey confirm the Copernican cause and the new physics. The Earth is not the single centre of gravity in the universe, and rotates on its own axis:

> When we rested . . . either we were insensibly carried (for I perceived no such motion) round about the Globe of the Earth, or else that (according to the late opinion of Copernicus) the Earth is carried about, and turneth round perpetually, from West to East, leaving unto the Planets onely that motion which Astronomers call naturall.[73]

The feeling of space flight is described by Godwin, as Gonsales gets progressively lighter on leaving Earth, and gradually heavier on approaching the Moon. Like

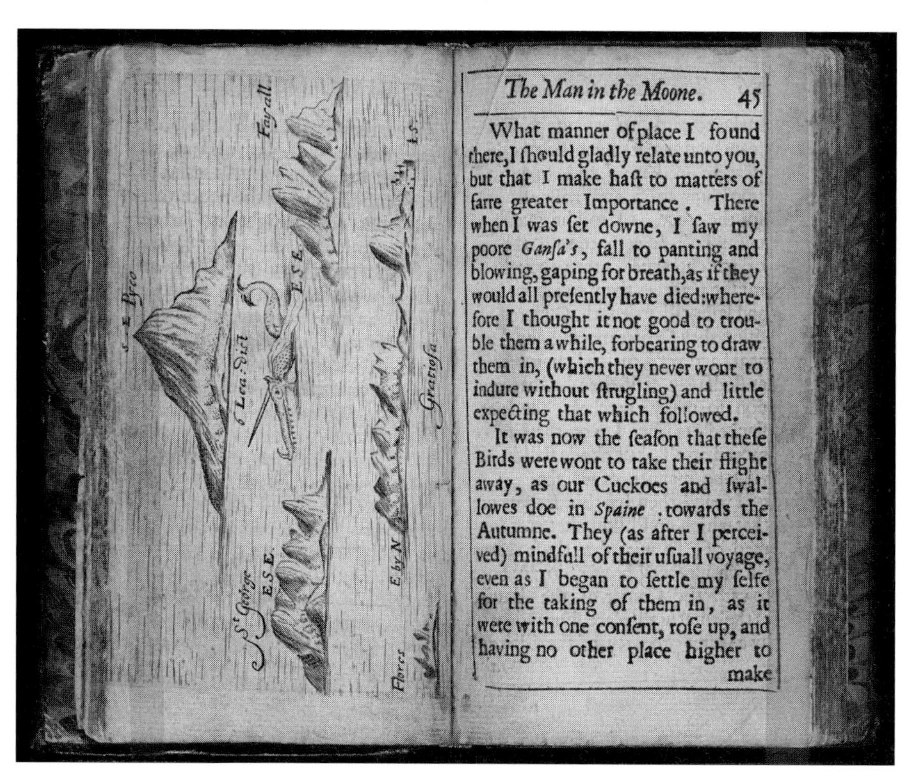

The Man in the Moone. 45

What manner of place I found there, I ſhould gladly relate unto you, but that I make haſt to matters of farre greater Importance. There when I was ſet downe, I ſaw my poore *Ganſa's*, fall to panting and blowing, gaping for breath, as if they would all preſently have died: wherefore I thought it not good to trouble them a while, forbearing to draw them in, (which they never wont to indure without ſtrugling) and little expecting that which followed.

It was now the ſeaſon that theſe Birds were wont to take their flight away, as our Cuckoes and ſwallowes doe in *Spaine*, towards the Autumne. They (as after I perceived) mindfull of their uſuall voyage, even as I began to ſettle my ſelfe for the taking of them in, as it were with one conſent, roſe up, and having no other place higher to make

Figure 7 To The Moon, By Goose! *The Man in the Moone: or, A Discourse of a Voyage Thither by Domingo Gonsales the Speedy Messenger*, by Francis Godwin.

Kepler, Godwin is flirting with gravity, an idea that had, in fiction and in fact, finally come of age. After a dozen days of flight, Señor Gonsales presently sets down on a lunar hill.

The Moon is a world like our own, but on a grander scale. Plants and animals all reach monstrous proportions, as do the lunar inhabitants. To be sure, in this Moon world, stature is a sign of nobility. So, 'true lunars' are thirty times taller than humans, living up to thirty times longer. Theirs is an idyllic life. In stark contrast, the 'dwarf lunars' are little taller than humans, and live no longer. They are assigned the most menial tasks. Given this lunar hierarchy, Señor Gonsales senses himself somewhat inferior.

Though Godwin claims his Moon world a utopia, his description makes you doubt it. The 'dwarf lunars' may be more virtuous than their terrestrial cousins. True, the most mortal wound can be healed. It seems crime is unknown, and Moon women are so striking no man is adulterous. But soon, it is revealed that the 'true lunars' rule by a militant form of eugenics. Flaws are identified at birth. And though the lunars' policy is not to kill, nonetheless the newborn defects are shipped Earthwards, to North America!

Francis Godwin's *Man in the Moone* is a key early work of space fiction. The notion of life beyond Earth was inspirational and, thanks to Kepler and Galileo, the potential reality of alien life dawned for the first time. Indeed, with the benefit of over-eager hindsight, the idea was mistakenly credited to Copernicus.

The good bishop imagined happy aliens. Godwin ventured that extraterrestrials were not only superior to us, a leaning with a long future history, they were also happier. Kepler's *Somnium* had imagined a nightmare alien existence, compared with life on Earth. Reincarnating the alien contact story, Godwin's is an early form of evolutionary fable. The cosmos may harbour a more highly evolved race than man. It was an idea that would become commonplace in fiction in the wake of Wallace and Darwin, over two hundred years later.

Fiction and the new physics

Space fiction also played a growing part in the popularisation of the new physics. Eventually, the evidence Galileo gleaned with the telescope made its mark. His infamous trial lent much esteem to the new materialist philosophy. But before the revolution in ideas could be played out in practice, its promise had to impact on those in positions of power: the new class of adventurers, politicians, and merchants.

The flow of information was vital. The cultural revolution in ideas sparked a sharp rise in Renaissance literacy, keenly among the middle classes, but also among the peasantry. The creation of the Gutenberg Galaxy led to the dawning of the age of mass communication. And so the gradual democratisation of knowledge began, including the first of the modern media phenomena, such as the press.

Witness the change in the century that separates Copernicus from Kepler. Copernicus had been so petrified by the likely religious reaction to *De Revolutionibus* that he had deferred publication until he lay on his deathbed. And even then the book's preface had been pressed with the daunting caveat '*For Mathematicians Only*'. Not exactly a battle cry. And no wonder the work became known as 'the book that nobody read'. In great contrast we have the contemporary imaginative writing, characterised by Galileo, and Kepler.

Galileo's little book *The Starry Messenger* struck like a bomb. It was the first time that anyone had provided strong visual evidence in support of the Copernican theory. Until then, the theory had seemed to challenge common sense and perception. As we have seen, the impact of Galileo's telescopic discoveries was utter shock. Even so, Galileo's book had the dramatic effect for which it was designed. It aroused fervent controversy.

The telescope inspired the imagination, and that spelt instant trouble.

According to the Aristotelians, which included the philosophers at the universities and the scholars of the Church, all that could be said about the heavenly bodies was that they were eternal, immutable, and perfect. They were, in short, fundamentally different from anything on Earth. And so it was prohibited to make any speculations about the Moon, or any other body, based on visual evidence. If anything was to be said about space, it was the job of the old philosophers, and certainly not the job of the new astronomers. Astronomy held a far inferior position in the hierarchy of disciplines, compared with theology, and was associated with mere numbers.

Space fiction threw these conventions into the dustbin of history. Orthodox histories of science have got it all wrong. They place Kepler in their timelines solely due to his innovation of the laws of planetary motion. But it is *Somnium* that gives us the true essence of Kepler. New worlds open to depiction are known as the *novum*. The *novum* cannot be understood without inventing stories through which we are better able to understand. The speculative space fiction, starting with Kepler, tried to slake the new desire to comprehend the new cosmos, more than just the mere data of the discovery. It promised to be communicable and rational, not the fruit of religious revelation. It was critical and reasoning, as well as creative.[74]

The view through the telescope was also limited. The far-seer's evidence was confined to the visual, and that made the discovery of the cosmos very different from most other discoveries of the age, in the far-flung parts of uncharted Earth. These new terrestrial territories were also 'unknown', up until the very moment of imperial 'discovery'. But they then became reachable, a relatively easy destination by ship.

The Moon, planets, and stars, however, remained beyond reach. The spyglass, revolutionary though it was, did not bring the cosmos any closer. Indeed, it made the imagined journeys even more fantastic. Terrestrial travellers might return from exploring distant lands with physical evidence of animals, or plants, or even another human being. At worst they would come back with eyewitness and mouth-watering tales, amazing stories of piracy and plunder, tales of creatures with an arm for a nose.

In contrast, consider the astronomer as explorer. True, the telescope also takes the traveller on a journey. But the exploration of the sky had to rely upon eyes only. In his *Starry Messenger*, Galileo had to convince his readers that his interpretation of what he saw through the spyglass was correct. And the almost fictional style he adopts in his book is a clear attempt to overcome the vast distance, especially by contemporary standards, between the Earth, the Moon, and the heavenly hosts. Galileo tried to transform the visual nearness, through the telescope, into a physical nearness, through his words.

The view through the spyglass was revolutionary indeed. And what the telescope saw was open to debate. It may have shown mountains and craters to some, but by no means to all. It was a visual experience starkly different from anything the human eye had seen before. So it was the job of writers like Galileo to convince the public of their interpretation of what the far-seer saw. It was down to their use of words to show similarity between Earth and Moon, and to show that what looked strange through the 'scope was not actually so. The readers needed to swap their strange experience for familiar ones, to ground their understanding, to bring the Moon down to Earth.

And yet the telescope implied so much more than you could actually see. It was limited in its visual field, while simultaneously suggesting a limitless number of new discoveries. Galileo did not set out his full lunar theory in a Copernican context until he published *Dialogue Concerning the Two Chief World Systems* (*Dialogo sopra i due massimi sistemi del mondo*), around twenty years after *The Starry Messenger*. It makes a fascinating read.

Galileo is at pains to avoid the alien. Both in the *Dialogue* and *Starry Messenger*, he is evasive when it comes to the question of what life might *look* like and *be* like, up there in the great beyond. He shuns fantastical digression and limits himself to justifiable interpretations of his view through the 'scope. In *Dialogue*, the discussion turns to whether the Moon houses plants, animals, or even humans, similar to those on Earth.

Though he is aware of the alien, and the potential of the imagination, Galileo does not choose to address such issues. Instead, reflecting on the probable wealth of diversity and difference in the new universe, he muses that things on the Moon may be so fundamentally different from things on Earth that they are entirely beyond the imagination. Consider, he says, 'a person born and raised in the forest among wild beasts and birds, and knowing nothing of the watery element'.[75] Such a person, Galileo argues, would never, not even with the most fertile imagination, be capable of imagining the ocean, with its boundless beasts, and shoreline domains and dwellings. He concludes, 'it is indeed possible to discover some things that do not and cannot exist on the Moon, but none which I believe can be and are there, except very generally.'[76]

As a gifted writer, as well as a gifted scientist, Galileo was not blind to the potency of prose. To begin with, he personifies the Moon and the Earth, in an attempt to bring them to life. For example, he uses the folklorist image of the face of the Moon, *facies Lunae*. This is a clear reference to *De facie in orbe lunae*, the work by first-century Greek Platonist Plutarch, which was a dialogue on different theories about the nature of the Moon. And throughout the *Dialogue* and *Starry Messenger*, Galileo uses terrestrial imagery that caters to the reader's imagination, and helps build his extraterrestrial and Copernican case.

In fact, Galileo uses the theory of *simulacra*. This theory of the 'images of things' was developed by the Atomist Lucretius in his book *On The Nature of Things (De natura rerum)*. Lucretius describes *simulacra* as subtle shells, or membranes, that spring from the surface of objects in order to roam about in space maintaining their forms.[77] For Lucretius, *simulacra* explain all phenomena relating to sensation, imagination, and memory.[78] And it is this technique that Galileo uses in feeding the readers' imaginations to enable them to visualise the strange and alien landscapes of the Moon and beyond.

One final word on Galileo. His example inspired a pluralist plea in 1622 from a Naples prison. There, Italian poet-theologian Tommaso Campanella wrote his *Apologia pro Galileo*. More than a decade earlier, Campanella had written to Galileo urging him to speculate on planetary dwellers. Now, his essay defends Galileo, the Copernican system, and the idea of other inhabited worlds. Campanella's aim was to save Galileo from the charge that he supported pluralism; though given the nature of this excerpt, it may have done more harm than good:

> If the inhabitants which may be in other stars are men, they did not originate from Adam and are not infected by his sin. Nor do these inhabitants need redemption, unless they have committed some other sin ... In his Epistle on the solar spots Galileo expressly denies that men can exist in other stars ..., but affirms that beings of a higher nature can

exist there. Their nature is similar to ours, but is not the same, despite whatever sportful and jocose things Kepler says at such length in his prefatory dissertation to the *Starry Messenger*.[79]

Beyond the new physics

Consider now the case of Kepler's fiction.

It is a cunning brand of Copernicanism. Though the empirical data and actual *proof* of the Copernican theory may have been lacking, Kepler effectively *invents* the proof, in the form of fiction. He writes a story about what the cosmos looks like from the lunar surface. The fictional approach allows assertions about the Moon's astronomy and topography, while the Kepler of the notes grins mischievously at his audacity in using fiction to force the physics. The astronomical thesis of *Somnium* is presented as a travel tale, an account of an actual journey. And the painstaking description of a journey to the Moon by fellow humans makes it all the more real.

As we discussed earlier in this chapter, though the goal of *Somnium* is scientific in nature, Kepler presents his thesis in the form of fiction, relying upon myth, legend, and folklore to make his mark. When it was published posthumously in 1634, *Somnium* comprised the brief but fantastic tale, along with Kepler's astronomy, geography, and biology of the Moon, and a series of explanatory notes. This commentary includes a letter, from Kepler, to the Jesuit mathematician Paul Guldin, *Geographical, or, if You Prefer, Selenographical Appendix*,[80] written in 1623.

Here, separated from the dream narrative, Kepler speculates on lunar fortifications. His thoughts on forms of life on the near side of the lunar globe (the Subvolvan hemisphere) are also detailed in the appendix, based on observations he made with the telescope, also in 1623. According to Kepler, the circular craters on the lunar surface, first discovered by Galileo, are circular fortresses, built by the rational Subvolvan civilisation. Indeed, he even makes some suggestions as to how the fortifications are built. Kepler then goes on to lay out his reasons for the fortification theory, as well as including his own translation, into Latin, of Plutarch's *De facie in orbe lunae*.

Kepler believed Moon people had made fortified cities. Galileo had speculated in his *Starry Messenger* that such spots, or craters, were cavities. His explanation of a particularly large cavity compared its topography to the geography of Bavaria. Kepler, however, was struck by the sheer number of such cavities. Given their round structure and their regular arrangement on the lunar surface, Kepler suggested they were artificial, not natural. In short, they had been made by the Endymionides, the rational inhabitants of the Moon.

The craters acted as fortified cities to shield them against the lunar climate and hostile forces. Kepler even suggests a way in which the Endymionides might measure out the space for their future cities, using a primitive type of compass:

> They drive down in the centre of the space to be fortified. To this stake they tie ropes, which are either long or short, depending on the size of the future town . . . With this rope fastened in this way, they move out to the future rampart's circumference, as defined by the ends of the ropes.[81]

The approach Kepler took was materialist and rational. He considered the lunar craters too perfect in shape to spring from a natural cause. They bore resemblance to medieval cities and 'are artificial and produced by some architectural mind'.[82] His interpretation was based on the idea that, for natural phenomena, 'the mind is the source of orderliness'[83] and that 'it therefore follows that those things which are out of order, insofar as they are of order, are the result of the movement of the elements and the composition of matter'.[84] On the other hand, 'when things are in order, if the cause of the orderliness cannot be deduced from the motion of the elements or from the composition of matter, it is quite probably a cause possessing a mind.'[85]

Once more, in fiction and in fact, the Moon was a touchstone.

If the two worlds of the Earth and the Moon can be linked through the principle of elemental matter in motion, so too can the rest of the cosmos. In *Somnium*, Kepler introduces this principle in relation to the average size of lunar creatures. Since Moon mountains are higher than earthly ones, he assumes Moon creatures too were much larger, 'This proportion is applicable not only to their bodies in comparison with our terrestrial bodies but also to their functions, breathing, hunger, thirst, waking, sleep, work, and rest.'[86] This idea links the topographies and lifeforms on the Earth and Moon both to their respective motions in space, and to cosmic perspective afforded by their locations:

> Between the very slow motion of the fixed stars for us and the brief periods of the individual planets down to the daily rotation of the Earth there is a proportion which seemed to me to be also that of human life to the modest size of our bodies. For the Moon, on the other hand, the fixed stars return more quickly than Saturn, whereas a day is thirty times longer than ours. Hence I thought that I should attribute a short life to the living creatures but enormous growth, so that nothing would attain a stable state and everything would perish in the midst of its development.[87]

The cosmic pluralism of Kepler is based on the mathematics of proportion. The various worlds in his cosmos are set in strict proportion with one another. Kepler's is a kind of Darwinian approach to cosmic evolution, the kind that cosmologist Lee Smolin was to develop in his 1997 book *The Life of the Cosmos*. For Smolin and his fecund universes theory,[88] it was the application of the principle of natural selection that bore baby universes. Hence, 'daughter' universes would have fundamental constants and parameters similar to that of the parent universe, though with some changes due to inheritance and mutation, as required by natural selection.

Using fiction and the Moon as his touchstones, Kepler told a similar tale. His pluralism is the logical extension of the principle of environmental adaptation that he dreamt up for the Moon. Kepler takes it as a fundamental creative law of the entire cosmos. It not only accounts for a cosmos of aesthetic harmony (like Pythagoras before him, Kepler was big on the harmony of the spheres), but also for the portrayal of planetary life, beginning with his lunar landscape.

The Moon was a stepping-stone.

Once the Earth was established as another planet in orbit about the Sun, the matter of Earth's relationship to the other planets followed logically. Galileo and others had used the spyglass to unveil the Earth-like nature of the Moon, at least in the non-Platonic sense of material mutability. Then there was the question of life on the Moon, or more specifically, rational life. Kepler had addressed some of these issues through the narrative of *Somnium*, in terms of his Endymionides, the rational inhabitants of the Moon.

But deep space remained unexplored. More moderate Copernicans such as Kepler limited their speculations to the planets circling the Sun. Soon, however, in fact and in fiction, deep space was ripe for speculation. In the second half of the seventeenth century it became common to understand the fixed stars of deep space as nothing other than suns in their own right, each with its own planetary system. And so the plurality of worlds became an infinity of worlds, and the cosmos grew in the minds of men.

Notes

1 Kepler, J. (1634) *Somnium*, trans. E. Rosen, New York, reprint 2003

2 Bernal, J. D. (1965) *Science in History, Vol. I*, London

3 Brecht, B. (1960) *Life of Galileo*, trans. D. I. Vesey, London

4 Hakluyt Society (1930) *Select Documents Illustrating the Four Voyages of Columbus*, ed. C. Jane, series II, vol. 65, London

5 Taylor, E. G. R. (1954) *Late Tudor and Early Stuart Geography 1583–1650*, London

6 Fernel was the first man of modern times to measure a degree of the meridian.

7 Sherrington, C. S. (1946) *The Endeavours of Jean Fernel*, Cambridge

8 Kuhn, T. (1957) *The Copernican Revolution: Planetary Astronomy in the Development of Western Thought*, Harvard

9 ibid.
10 Milton, J. (1674) *Paradise Lost* (Second Edition), London
11 Koestler, A. (1959) *The Sleepwalkers*, Baltimore
12 Brake, M. and Hook, N. (2007) *Different Engines: How Science Drives Fiction and Fiction Drives Science*, London
13 Kuhn, T. (1957) *The Copernican Revolution: Planetary Astronomy in the Development of Western Thought*, Harvard
14 Grandgent, C. (1924) *Discourses on Dante*, Harvard
15 Bernal, J. D. (1965) *Science in History, Vol. I*, London
16 Brake, M. and Hook, N. (2007) *Different Engines: How Science Drives Fiction and Fiction Drives Science*, London
17 Bernal, J. D. (1965) *Science in History, Vol. II*, London
18 Butterfield, H. (1957) *The Origins of Modern Science*, London
19 Royal Astronomical Society (1947) *Nicholas Copernicus, De Revolutionibus, Preface and Book I*, trans. J. P. Dobson, S. Brodetsky, *Occasional Notes*, no. 10
20 ibid.
21 Koyré, A. (1961) *La Révolution Astronomique*, Paris
22 Kuhn, T. S. (1957) *The Copernican Revolution*, Harvard
23 Koestler, A. (1959) *The Sleepwalkers*, London
24 Royal Astronomical Society (1947) *Nicolaus Copernicus, De Revolutionibus, Preface and Book I*, trans. J. P. Dobson, *Occasional Notes*, no. 10
25 Dick, S. J. (1998) *Life on Other Worlds*, Cambridge
26 Bernal, J. D. (1965) *Science in History, Vol. II*, London
27 Galileo Galilei, *Dialogue Concerning the Two Chief World Systems*, trans. Stillman Drake (1967), Berkeley
28 St Augustine, *Works*, ed. M. Dods (1870), Edinburgh
29 Oldridge, D. (2007) *Strange Histories*, Oxford
30 Weiner, A. D. (1980) 'Expelling the beast: Bruno's adventures in England', in *Modern Philology*, **78**, no. 1, pp. 1–13
31 Dick, S. J. (1996) *The Biological Universe*, Cambridge
32 Ingersoll, R. G. (1902) *The Collected Works of Robert G. Ingersoll*, New York
33 *Il Sommario del Processo di Giordano Bruno, con appendice di Documenti sull'eresia e l'inquisizione a Modena nel secolo XVI*, ed. Angelo Mercati, in *Studi e Testi*, vol.101
34 Foote, G. W. and McLaren, A. D. (1886) *Infidel Death-beds*, London
35 ibid.
36 Rose, M. (1982) *Alien Encounters: Anatomy of Science Fiction*, Harvard
37 Kessler, J. J. (1997) *Giordano Bruno*, Stark
38 Rabin, S. (2005) 'Nicholaus Copernicus', in *Stanford Encyclopaedia of Philosophy*
39 Turner, W. (1908) 'Giordano Bruno', in *Catholic Encyclopaedia*
40 http://asv.vatican.va/en/doc/1597.htm
41 Ingersoll, R. G. (1902) *The Collected Works of Robert G. Ingersoll*, New York
42 Yates, F. (1964) *Giordano Bruno and the Hermetic Tradition*, London
43 Lambert, L. B. (2002) *Imagining the Unimaginable: The Poetics of Early Modern Astronomy*, Amsterdam

44 Galilei, G. (1610) *Sidereus Nuncius*, Venice
45 George Fugger in a letter to Johannes Kepler, 16 April 1610, *Gesammelte Werke*, vol. XVI, p. 302
46 Galilei, G. (1610) *Sidereus Nuncius*, Venice
47 ibid.
48 ibid.
49 Drake, S. (1957) *Discoveries and Opinions of Galileo*, New York
50 Hallyn, F. (1990) *The Poetic Structure of the World: Copernicus and Kepler*, New York
51 Lambert, L. B. (2002) *Imagining the Unimaginable: The Poetics of Early Modern Astronomy*, Amsterdam
52 Koestler, A. (1959) *The Sleepwalkers*, London
53 Rosen, E. (1967) *Kepler's Somnium*, Madison and London
54 Christianson, G. (1976) Kepler's *Somnium*: Science Fiction and the Renaissance Scientist, *Science Fiction Studies*, Indiana
55 Caspar, M. (1959) *Kepler*, New York and London
56 *Gesammelte Werke*, ed. W. Von Dyck, M. Caspar *et al.* (1937) vol. 11, pt. 2, p. 330, Munich
57 Koestler, A. (1959) *The Sleepwalkers*, London
58 *Gesammelte Werke*, ed. W. Von Dyck, M. Caspar *et al.* (1937) vol. 11, pt. 2, p. 330, Munich
59 ibid.
60 ibid.
61 ibid.
62 ibid.
63 Cola is part of an old legend of a young sea diver who, like a fish, used to hide under water for three or four hours. A fact that Kepler's reverence for his sources prevented him from realising.
64 *Gesammelte Werke*, ed. W. Von Dyck, M. Caspar *et al.* (1937) vol. 11, pt. 2, 330, Munich
65 ibid.
66 Nicolson, M. H. (1948) *Voyages to the Moon*, New York
67 Brake, M. (2003) *On the Plurality of Inhabited Worlds: A Brief History of Extraterrestrialism*, Wales
68 Parrinder, P. (2000) *Learning from Other Worlds: Estrangement, Cognition, and the Politics of Science Fiction and Utopia*, Liverpool
69 Dick, S. J. (1996) *The Biological Universe*, Cambridge
70 Rose, M. (1982) *Alien Encounters: Anatomy of Science Fiction*, Harvard
71 Wilkins, J. (1640) *The Discovery of a World in the Moone*, London
72 ibid.
73 Godwin, F. (1638) *The Man in the Moone*, London
74 Brake, M. and Hook, N. (2007) *Different Engines: How Science Drives Fiction and Fiction Drives Science*, London
75 *Le Opere di Galileo Galilei* (1890–1909) Vol. 7, 86, ed. A. Favaro, Florence
76 ibid.
77 Lambert, L. B. (2002) *Imagining the Unimaginable: The Poetics of Early Modern Astronomy*, Amsterdam
78 ibid.
79 Campanella, T. (1937) The Defence of Galileo, trans. G. McColley, *Smith College Studies in History*, **22**, nos. 3 and 4, pp. 66–7

80 Appendix Geographica, seu mavis, Selenographica, *Gesammelte Werke*, ed. W. Von Dyck, M. Caspar *et al.* (1937) vol. 11, pt. 2, p. 368, Munich

81 ibid.

82 ibid.

83 ibid.

84 ibid.

85 ibid.

86 ibid.

87 ibid.

88 The fecund universes theory, also known as cosmological natural selection theory, advanced by Lee Smolin suggests that a process analogous to biological natural selection applies at the grandest scales.

3

In Newton's train: pluralism and the system of the world

'Sir,' I replied to him, 'the majority of men, who only judge things by their senses, have allowed themselves to be persuaded by their eyes, and just as the man on board a ship which hugs the coastline believes that he is motionless and the shore is moving, so have men, revolving with the Earth about the sky, believed that it was the sky itself which revolved about them. Added to this there is the intolerable pride of human beings, who are convinced that nature was only made for them – as if it were likely that the Sun, a vast body four hundred and thirty-four times greater than the Earth, should only have been set ablaze in order to ripen their medlars and to make their cabbages grow heads!'

'As for me, far from agreeing with their impudence, I believe that the planets are worlds surrounding the Sun and the fixed stars are also suns with planets surrounding them; that is to say, worlds which we cannot see from here, on account of their smallness, and because their light, being borrowed, cannot reach us. For how, in good faith, can one imagine these globes of such magnitude to be nothing but great desert countries, while ours, simply because we, a handful of vainglorious ruffians are crawling about on it, has been made to command all the others? What! Just because the Sun charts our days and years for us, does that mean to say it was only made to stop us banging our heads against the walls? No, no, if this visible god lights man's way it is by accident, as the King's torch accidentally gives light to the passing street-porter.'

Cyrano de Bergerac, *Les États et Empires de la Lune*, trans. G. Strachan[1]

The System of The World

The Scientific Revolution did not take place in a vacuum.

Copernicus, or Kepler, or Galileo, did not just wake up one morning, as conventional histories of science would have us believe, and declare their discoveries to an ignorant world. As American philosopher John Herman Randall said on the matter,

> One gathers, indeed, from our standard histories of the sciences, written mostly in the last generation, that the world lay steeped in the darkness and night of superstition, till one day Copernicus bravely cast aside the errors of his fellows, looked at the heavens and observed nature, the first man since the Greeks to do so, and discovered… the truth about the solar system. The next day, so to speak, Galileo climbed the leaning Tower of Pisa, dropped down his weights, and as they thudded to the ground, Aristotle was crushed to earth and the laws of falling bodies sprang into being.[2]

Such ideas of scientific progress ignore two crucial factors. The first factor is the long periods of investigations, which rely upon a gradual advance on tradition and custom, and is the fruit of many ordinary thinkers and workers. The second factor is the revolutionary tipping points. In spite of all gradualness, an innovatory leap leads to decisive change, with such leaps being associated with the 'great men of science'.

There is a problem with this 'great men' myth. It has led to a false idea of science, an idea which suggests that progress in science is due solely to the genius of great men, irrespective of factors such as culture, society, and economy. We are asked to believe that these masterminds just dream this stuff up out of thin air. Many conservative histories of science are rooted in the 'great men' myth. They are little more than a series of naïve narratives of great discoverers, each with their own momentous and revelatory insight into the secrets of nature.

For progress in science is not solely due to the genius of great men. Even if that man is Isaac Newton. The biographer Richard Westfall said about Newton that the more he studied him, the more Newton receded from him. Westfall had known a number of brilliant men, but none quite like Newton. In fact, Westfall went as far as to say that Newton 'has become for me wholly other, one of the tiny handful of supreme geniuses who have shaped the categories of the human intellect'.[3]

You can see the problem. To be clear, great men such as Galileo or Newton have been a vital influence in the development of science. But their contribution should be studied in context, and not one step removed from their social setting. An inability to see this often leads to the use of redundant words like 'brainwave' or 'genius' to explain away those eureka moments of discovery.

In fact, conventional histories devalue great men. Such accounts are too narrow and too idle to realise that great men were creatures of the culture in which they swam. Once we recognise that, we understand that such men were subject to the same sway of social influence, the same sorry compulsions, and their stature is enhanced. Indeed, as we shall see, the greater the man, the more he was immersed in the milieu of his days. He becomes more important. For only by seizing the moment chanced by his times was he able to make that innovatory leap that led to critical change.

Now Newton's 'system of the world' is a case in point. *Mundi Systemate* (*On the System of the World*) was the third book of his magnum opus, *Philosophiæ Naturalis Principia Mathematica* (*Mathematical Principles of Natural Philosophy*), often known as simply *Principia*, and first published 5 July 1687. *Mundi Systemate* was an explanation of the consequences of his theory of gravity, but on a cosmic scale. The first theory of everything, if you like.

The idea of a 'system of the world' was a turning point for the Scientific Revolution. As the Platonic–Christian view of the world began to crumble, a new vision emerged. Using the medieval roots of science, derived from the Greeks through Islamic scholars, the revolutionaries created a new view of the cosmos, nature, and, ultimately, man.

The revolution had a backstory. The body of ancient Greek beliefs, including most notably for us Atomism, had been refined and filtered by Islamic scholars in the eleventh and twelfth centuries. It was to be found at the school of Chartres in the mid twelfth century, and at the medical school at Salerno near Naples in 1060. In Toledo, around a hundred Arabic works had been translated, along with Ptolemy, by 1175. Also by the twelfth century, Islamic natural philosophy and mathematics had found its way to Oxford, and to Padua. And so from the early twelfth century, we can see a golden thread, the development of a nascent and alternative worldview. Though the likes of Atomism may have been overshadowed by the intellectual bulk of Aristotle, the tradition lived on, through the fifteen and sixteenth centuries, and well into the seventeenth.

Such was the long road to modern science. Spun out of the fabric of history, and not out of empty space. The revolution in science developed over time. As the Church continued to wield absolute power, authority, and obedience, the emerging counter-culture of science lurked for five centuries preceding the watershed of the Scientific Revolution.

And yet there were crucial tipping points. One of the major innovative leaps, which led to significant change, was the work of seventeenth-century French philosopher, René Descartes, perhaps best known for the philosophical statement 'Cogito ergo sum' ('I think, therefore I am'). Descartes' book, *The World*, originally titled *Le Monde* and also called *Treatise on the Light*, is an early attempt at a theory of everything. Written between 1629 and 1633, when Descartes was in his early thirties, the book contains an almost complete version of his philosophy, including method and metaphysics, physics and biology.

Descartes was an Atomist and a materialist. His materialism took the form of a mechanical philosophy, which held that the cosmos is best understood as a mechanical system, a universe composed entirely of matter in motion, and under the sway of laws of nature. Descartes' Atomism is often described, rather wordily, as Corpuscularianism. This viewpoint, closely linked to Atomism in most regards, is a belief that everything in the physical universe is made of tiny 'corpuscles' of matter. Indeed, Descartes' only main disagreement with Atomism per se was that he thought there could be no vacuum.

Descartes was canny. *Le Monde* endorsed a Copernican view of the cosmos. As Descartes saw it, his cosmology of constantly swirling matter explained, among other things, the creation of our Solar System, and the motion of the planets about the Sun. But he had seen what had become of Galileo when the Catholic Church dealt with those who publicly aired such beliefs. Descartes delayed the release of *Le Monde*, only to find his influence making its mark on other writers and thinkers of Renaissance Europe. So let us continue our journey, as the seventeenth century bleeds into the eighteenth, and the Cartesian system of the world find its reflection in fiction.

The long road to cosmic pluralism

Copernicanism caught fire.

Seventeenth-century thinkers grasped the idea of plural worlds, not just as a plurality of worlds similar to Earth in topography, but as a plurality of inhabited worlds, peopled moreover by rational beings. Naturally, the supposed presence of extraterrestrial intelligence (ETI) was the most useful tool of comparison between Earth and other potential worlds. So, there is little doubt, as Giordano Bruno's persecution witnessed, that the Copernican worldview contained a golden thread of cosmic pluralism running through it.

Not that Copernicus himself imagined a universe full of aliens. Far from it. But the Earth's demotion from cosmos-central to that of mere planet certainly had profound consequences. For one thing, the Earth's downgrading was also charged with the principle of plenitude. Like much of cosmology, the idea of plenitude had been present for some time in the two main traditions, idealist and materialist. In idealism it had been present in Platonic geocentrism, the notion of full creative power realised through a strictly vertical cosmic order and theological hierarchy.

Atomism too had its principle of plenitude. This materialist form of the belief assigned the creative power to Nature, and not to God. The idea of 'fullness' was now transformed into fecundity, 'when abundant matter is to hand, and no thing and no cause hinders, things must assuredly be done and completed'.[4] And so the strict hierarchy of the Platonic–Christian principle became a levelled and materialist one.

The Atomists brought another vital idea to the debate; and that was the principle of mediocrity. This idea is a belief in the uniformity of nature, the notion that nothing is unique in the universe, but that everything belongs to some generic type, or class. The proposal was powerful in its simplicity. Was not the Earth now just another planet in a dance about the Sun? Might it not also be true that other planets, or indeed other Earths dance about other stars in the same way? And are there not multitudes of such suns, swarming in deep space? For heresies such as these Giordano Bruno was burned at the stake, at the beginning of the seventeenth century. By the end of that century, Lucretius' *De Natura Rerum* (*On the Nature of Things*) had once again become very influential, and the materialist ideas of Atomism gifted a full recovery to the idea of plural worlds.

Empires of Moone and Sun

The Moon captured the Western imagination. Not only was it the subject of telescopic discovery, and doorstep witness of material similarity between Earth and cosmos, but readers may also have pored over these early space journeys by the light of the Moon itself. Unlike the far-flung destinations of the voyages of less earthly discovery, the Moon gazed down, clear for all to see.

The rocket made an early appearance.

Kepler had divined 'Moon spirits' for his lunar journey, while Godwin had favoured the wild goose chase. But in a single generation the space voyage quickly evolved into a rocket-propelled critical 'contact' fiction, pioneered by Cyrano de Bergerac. Cyrano was, of course, notoriously unorthodox. French satirist and freethinker, his was a life immortalised by romantic legend. But the real-life counterpart equalled the legend of the swashbuckling swordsman with a large snout. Cyrano allegedly fought with the elite Gascon Guard, and was a radical who liked to parade his heterodoxy in extravagant style.

He had studied with French philosopher Pierre Gassendi, whose life's work was to change the public perception of science. Gassendi was a canon of the Catholic Church who tried to reconcile Epicurean atomism with Christianity. He had a reputation as a successful communicator of science, a reputation that owed much to the influence of his famous pupil. And Cyrano's repute rested on a trilogy of tales, known collectively as *L'autre Monde* (*Other Worlds*), and separately as *Les États et Empires de la Lune* (*The States and Empires of the Moon*) (1657), *Les États et*

Empires du Soleil (*The States and Empires of the Sun*) (1662) and the lost work, *L'histoire des Etoiles* (*The History of the Stars*).

In his *Lune* and *Soleil*, Cyrano makes an Atomist argument for a plurality, if not an infinity, of inhabited worlds. As well as the influence of his mentor Gassendi, Cyrano was also swayed by Descartes. Descartes was famed for postulating the Atomist cosmological theory of vortices, which proposed that the cosmos was filled with matter in various states, swirling in vortices about stars like the Sun. Unlike Descartes, Cyrano also radically assumed the existence of as many Copernican planetary systems as there are suns. Though Cyrano was writing in fiction, rather than fact, his theories nonetheless compose a system of the world. The system is presented mostly in the form of conversations in Cyrano's narratives, where his protagonist encounters inhabitants of both Moon and Sun, in the most imaginative and mischievous tales.

His imagination captured the mood of the age. Like Lucian, Cyrano's imagined alien life was openly atheist. Copernicus, Galileo, and Kepler had been pious men. Not Cyrano. Bishops Godwin and Wilkins had imagined cosmic life as further proof of God's glory. Under the influence of Descartes, Cyrano banished God from the heavens, with pure reason as his guide. His alien tales were remarkably popular all over Europe, and an influence on writers such as Jonathan Swift, Edgar Allan Poe, and probably Voltaire.

Cyrano took great delight in scandalising the Church. The first two volumes of his trilogy were so strident that they had to be toned down in their sacrilege and heresy. The secular culture implied by the new physics was quite openly expressed in Cyrano's space fiction. The notion of a cosmos fit for life was purged of any divine nuance. Cyrano's alien worlds were not the work of some Creator. And his space fiction was a unique attack on human self-esteem, and hubris. Kepler had confronted the Copernican cosmos in fiction. Now, and with comparable daring, Cyrano tackled the topic of Cartesian physics, a non-Christian view of an alien universe, which Descartes himself shied away from in public.

Books one and two of *L'autre Monde* told tales of imagined meetings between man and alien. Kepler had presented a Moon world in *Somnium*, which was at least in some ways based upon observation and number. Cyrano does no such thing. Cyrano's fictions of the lunar and solar worlds make little pretence that they are based on observational reality. Rather, they are fictitious, off-world explorations of arrogant and narrow-minded objections to pluralism.

The cosmos of *L'autre Monde* is Atomist.

One of the characters we encounter on Cyrano's Moon explains, 'From a serious investigation of matter, therefore, you would discover that it is all one, but that, like an excellent actor, it plays all kinds of roles in this life, in all kinds of disguises.'[5] Should there be some objection to this, or some 'recourse to the idea of Creation',[6] Cyrano explains further,

> Yet you are still astonished at the way this matter, mixed up pell-mell at the whim of chance, could have produced a man, seeing how many things were necessary for the construction of his person. Are you not aware that this matter has stopped a million times on its way towards the formation of a man, sometimes to make a stone, sometimes a lump of lead, sometimes coral, sometimes a flower, sometimes a comet? All this happened because there were more or less of certain shapes, which were necessary, or certain shapes, which were superfluous to the design of man. Hence it is no marvel that they should have come together, from among an infinity of substances which are shifting and changing incessantly, to make the few animals, vegetables, and minerals which we see, any more than it is a marvel for a triple number to come up in a hundred throws of the dice, since it is impossible for this movement not to produce something. And a fool will always marvel at this thing, not knowing how near it came to not being made.[7]

A cosmos is conjured up, one in which there are as many planets as stars, and these numberless worlds, like the Moon and Sun in his stories, are all peopled by rational creatures. An intelligent universe is only the beginning. Cyrano uses this cosmic landscape to cast doubt over the nature of man. His conclusions are not reassuring. The books are witness to the most critical view of the alleged superiority of man, and of religion, than any other work of its time.

The Empire of the Moone

Cyrano's voice is crystal clear in *The States and Empires of the Moon*. For one thing, the narrator is named Dyrcona, an anagram of d[e] Cyrano. And for another, what unfolds is nothing short of a lunar lampoon of hubris on Earth. The narrative begins, ingeniously enough, with space travel by bottled dew!

Take one: by fastening the bottles about his waist, the evaporating dew lifts Dyrcona to the Moon. But his mission fails, and he falls to Earth, initially to the French colony of Canada. Once captured, the Viceroy, who is very interested in Dyrcona's means of travel and the new astronomy, questions Dyrcona. He replies, 'I believe that the planets are worlds surrounding the sun and the fixed stars are also suns with planets surrounding them.'[8] The Viceroy becomes a Copernican and concludes that the cosmos is vast, and Aristotle wrong.

Figure 8 If the sun 'draws up' dewdrops then, suggested Cyrano de Bergerac, one might fly by trapping dew in bottles, strapping the bottles to oneself, and standing in sunlight.

Take two: by fixing fireworks to a travelling machine, Dyrcona now rockets to the Moon. In the opinion of twentieth-century science-fiction author Arthur C. Clarke,[9] Cyrano invented the ramjet, a form of air-breathing jet engine, which uses an engine's forward motion to compress incoming air. Though the invention of the ramjet is traditionally attributed to French inventor René Lorin in 1913, Clarke drew attention to this innovative passage by Cyrano, written around 250 years before Lorin:

> I foresaw very well, that the vacuity ... would, to fill up the space, attract a great abundance of air, whereby my box would be carried up; and that proportionable as I mounted, the rushing wind that should force it through the hole, could not rise to the roof, but that furiously penetrating the machine, it must needs force it upon high.[10]

The Moon that Cyrano unveils is another world. As in Godwin's tale, Cyrano's lunar landscape is stalked by a race of giants. It is a case of four legs good, two legs bad, as the lunar men, the Lunarians, walk on all fours, and greet Dyrcona with some disdain – a little bi-ped from their 'moon'. The lunar beasts too walk on all fours, making Dyrcona's appearance look all the more alien. On Cyrano's Moon, humans are not as fit as beasts, let alone the Lunarians. Dyrcona encounters a Phoenix and an extraterrestrial intelligence in the form of talking trees. He learns that rather than burn in his nest to birth a new bird, the Phoenix flies off to the Sun, to join the community of birds there; while the trees maintain that the reason and soul within them are their own, and not the result of humans morphed into trees. They are truly alien.

The medieval-minded Church suffers a lunar lampoon. Scriptural claims that only man has the faculties of reason and immortality are made to look ridiculous. After a dazzling description of an atheist genesis of life in the universe, and a committed declaration that the stars are inhabited, Cyrano explains that reason is the keystone of the enlightened lunar culture. In their culture, death is the fulfilment of life, and man's belief in God is considered a deficiency of reason. The decidedly intelligent Lunarians feed on scents, and sleep on flower blossoms.

Poetry is used as a form of tender. As if to graphically illustrate the richness of cultural life on the Moon, Dyrcona's guide on the Moon pays their bill in a pub using poetry. He outlines how on the Moon poetry serves as money:

> When some have been composed, the author brings them to the Court of the Mint, where the Jury of Poets of the Realm hold their sessions. There, these Officers of Versification put the pieces to the test and if they are judged to be of good alloy, they are taxed not according to their weight but according to their wit, ... Thus when someone starves to death it is never anyone but a blockhead and witty people always live off the fat of the land.[11]

Fiction soon catches up with fiction. Cyrano is very consciously playful about his Moon world. He has Dyrcona meet up with Domingo Gonsales, from Francis Godwin's *The Man in the Moone* tale. It is a curious juxtaposition. In one sense it gives Cyrano's tale gravitas, as Godwin's story was a very popular one. But the affirmation bestowed by the Spaniard's presence on the Moon is offset by the story's details. Dyrcona's description of the lunar world differs in most respects from that given in Godwin's novel.

Humanity's place on the cosmic ladder is also the rub in a run of weird events. The Lunarians train Dyrcona to perform tricks, just like chimps on Earth. He learns that the lunar men think Gonsales a baboon. The Spaniard is kept as a ludicrous pet, and shown in a lunar zoo. Indeed, the discussions that Dyrcona has with Gonsales while at the zoo are especially amusing to the Lunarians. But when the rumour starts that the two caged bi-peds may be flawed 'humans', the lunar church intervenes. A law is decreed: any suggestion of similarity between the earthly creatures and Lunarians, or even lunar beasts, is blasphemy.

Finally, Dyrcona stands trial. The charge against him: believing his world is central, and not a 'moon' of the Moon! The lunar men argue that humanity is completely insignificant. The bi-peds are considered monsters devoid of reason, best thought of as 'plucked parrots'. Paradoxically, the court still decides to declare Dyrcona a 'man', so he can be charged. His humiliation is to publicly recant his geocentric beliefs, on every street corner of the Moon.

The Empire of the Sun

Cyrano's obsession with the status of man continues in the second volume of his trilogy, *The States and Empires of the Sun*. This time, Dyrcona's preferred method of flight is the solar reflector, which he ingeniously uses to steer a flying machine sunward. The voyage lasts four months, after which Dyrcona sets down on a sunspot. Dyrcona's guide around the world in the Sun is Italian philosopher Tommaso Campanella, whose own book, *The City of the Sun*, describes a theocratic society where goods, women, and children are held in common. Dyrcona is allowed to witness things that no other human has experienced. In this way, he is entrusted with the duty of revealing to his fellow humans all about his solar journey.

Dyrcona's solar sojourn draws a distinction between the light and dark solar regions. For the most part, he travels the darker regions. Only in the first part of his journey does he find himself in the *lumineuse campagne* of the light solar regions. Here, he totally loses his orientation. It feels as though he is walking on nothing, neither is there anything his eyes can rest upon. He is forced to count his steps carefully, and finally feels relieved when he reaches the respite of the dark regions. Here, in the dark, his body seems secretly in step. At once he falls asleep, and on awaking finds a more familiar and inhabited world, fit with topographical features. Though many other solar creatures turn out to be more complex, the inhabitants of the darker solar regions all originate from the Earth, and their forms are terrestrial.

Cyrano flirts with an early form of panspermia. The topographical features that characterise the darker regions have been imported from the Earth. A forest of talking trees also has terrestrial origins, grown from seedlings of the mythical oaks of Dodona,[12] and carried to the Sun in the gut of an eagle, and spewed out undigested:

> A great eagle ... was feeding upon the acorns which their branches offered it ... Tired of living in a world where it had suffered so much, it took flight to the sun and made such a successful trip that it finally reached this luminous globe, where we are now. But on its arrival, the heat of the climate made it vomit and it brought up a number of still undigested acorns. These acorns germinated and from them there grew the oak trees which were our ancestors.[13]

Open-mouthed, Dyrcona observes the continuingly stunning sight of quick-time solar evolution: living creatures continuously morph into birds, trees, and human beings, until they are ultimately disclosed as beings more evolved than man. The Sun is unveiled as a strange but wonderful world of colour, a globe of 'burning snowflakes', and with plant life of stunning beauty. On waking at the foot of a tree all made of gold and adorned with emerald leaves, and buds of pearls and diamond flowers, Dyrcona meets with the king of a solar nation. Sitting on the tree is a nightingale. It sings the most exquisite melodies.

When Dyrcona spies a pomegranate of astonishing beauty, hanging on the tree, he is staggered when it morphs before his eyes into a dwarf, falling onto the ground next to him. The dwarf introduces himself as the king of a nation contained in the tree. Another metamorphosis sees the tree suddenly burst into a number of little men (not green). The nightingale alone remains, piping its melancholy love songs. When asked why the bird does not morph, the king replies, 'It is a true bird, which is no more than what it appears to you to be'.[14] The king then tells the story of the birds.

In the passage *Histoire des Oiseaux* (*Story of the Birds*), the Sun is revealed in all its magnificence. To better communicate his tale, the king turns into a midsized man. Now the king explains how he and his people are natives of the light region of the Sun, and that they spend much of their time travelling through the vast solar regions, studying the diverse customs of their inhabitants, their respective climates, and so on. The king explains that their ability to metamorphose is one outcome of their powerful imagination, which allows them an elastic ability with matter:

> You must know that being inhabitants of the pellucid part of this great world, where the principle of matter is to be in action, we are bound to

> have much more active imaginations than those of the opaque regions, and bodies composed of a much more fluid substance. Once one supposes this, it inevitably follows that our imagination, encountering no resistance in the matter of which we are composed, can arrange it at will, and, having become the complete master of our bodies, manoeuvres them, by moving all their particles, into the arrangements necessary to constitute the objects it has pictured in miniature. . . . You men cannot do such things on account of the heaviness of your substance and the frigidity of your imagination.[15]

Cyrano demonstrates his Atomist theory of everything.

For instance, the king and his people are a curious form of life, a fictionalised form of Atomistic theory. The morphed forms in which his people appear are merely recompositions of the same matter, evolution in fast forward. In Cyrano's fiction the detail of these material metamorphoses seem to serve a specific purpose. It presents an Atomist view of life in the universe, with the lifeforms defined as the composite of a great number of component parts. In the narrative, a mirror image of Dyrcona himself is conjured up, a microcosm of tiny solar organisms. Cyrano demonstrates a theory of the fundamental equality of matter, which underlies different phenomena. Despite appearances, and the way in which trees and pomegranates morph into people, matter remains the same, it is the agent of change.

The ultimate realm of the Sun is portrayed as the domain of the souls of all the creatures in the cosmos, a realm of truth and of peace, the province of lovers, and the state of philosophers. Mostly, however, it is the abode of the birds. Here, Dyrcona's solar honeymoon is fleeting as he is again put on trial. Dyrcona's capture by the birds enables Cyrano to develop the theme of man as a plucked bird. To the cultured birds, man is seen as a rather deluded creature. Man supposes the entire animal kingdom to be at his disposal, the environment too. Man is a feather-free monster, a creature of conflict, and a brutal killer. In a vain and devious attempt to throw his judges, Dyrcona makes believe he is an ape. But the birds are not to be fooled. They decide that man is a bane of existence, 'of which every well-policed state should be rid'.[16]

The bodies of the Moon and Sun in Cyrano's *L'autre Monde* are worlds like our own.

Cyrano's solar birds are as arrogant as humans. Like man, they believe themselves the supreme, most rational, and most cultured creatures in the cosmos. They are just as cruel. They try to condemn Dyrcona to the grisly fate of being eaten alive by flies, a fate he narrowly avoids. And Cyrano's Lunarians regard humans as circus monsters, suggesting humans might be guilty of the same cruelty, should the Moon dwellers ever come to Earth. The Lunarians delight in dressing Domingo Gonsales as a pet monkey. And Gonsales, with similar arrogance, tells the Moon dwellers that all life in the cosmos is created for Spaniards to enslave.

L'autre Monde is Cyrano's masterpiece. It is a biting satire of the bigotry and beliefs of the age. And it portrays a most secular view on the potential meaning of a universe fit for life. Cyrano's contribution to pluralism was an early Copernican stand, a declaration against the anthropocentric idea that man was the centre of creation. Space fiction played a key part in the early development of pluralism. The combined efforts of Johannes Kepler, Francis Godwin, and Cyrano de Bergerac contributed to a culture that meant every knowing cosmology since has held the prospect of alien life as a real possibility.

Talking alien

Lucretius' Atomist vision of the principle of plenitude lived on.

It was again influential, later in the seventeenth century, in the publication of the first work of pluralist science communication. *Entretiens sur la pluralité des mondes* (*Conversations on the Plurality of Worlds*), by French author Bernard le Bovier de Fontenelle, planted the idea of alien life even deeper into the European psyche. Fontenelle's book championed not only Copernicanism, but also pluralism, and is considered one of the first major works of the Age of Enlightenment.

The popularity of the book meant that it was translated into Danish, Dutch, German, Greek, Italian, Polish, Russian, Spanish, Swedish, and English. Indeed, by 1688 there were three different English translations, with five more following. In total, this incredibly popular volume has gone through about one hundred editions.[17] Unlike Cyrano, Fontenelle avoided challenging the Christian perspective, though the 1740 translation in Russian was censored, due to opposition from the Russian Orthodox Church. Although Fontenelle's book charmed the European public, the Roman Catholic Church also deemed it dangerous. *Conversations* was placed on the Index of Prohibited Books in 1687, removed in 1825, and bizarrely added again in 1900.[18] Once again, pluralism broke new barriers. Unlike most scientific treatises of the time, *Conversations* was written not in Latin, but in French. It was also an early example of a work that attempted to explain science in a popular way. In the book's preface, Fontenelle specifically addresses his female readers. And in a move that has since been praised by modern feminist critics, Fontenelle explains that his books are to be read by all (or at least by the chattering classes), and understood by those without an education in science.

Fontenelle made philosophy fashionable. Nephew of great French dramatists Pierre and Thomas Corneille, Fontenelle had initially trained in law, but gave up after just one case and devoted his life to scribing about philosophers and scientists, especially in defence of Atomism and the Cartesian tradition. Fontenelle wrote widely on the nature of the universe: 'Behold a universe so immense that I am lost in it. I no longer know where I am. I am just nothing at all. Our world is terrifying in its insignificance.'[19] *Conversations* became the most influential work on the plurality of worlds in the period, and was one of the main reasons Fontenelle was named Perpetual Secretary to the French Academy of Sciences, an office he held for forty-two years.

The text itself is presented, as the title suggests, as a series of conversations, between a gallant philosopher and a Marquise.[20] Together they walk in the noblewoman's gardens at night, gazing up at the starry skies of France. The philosopher explains Copernicus, and ponders on the possibility of extraterrestrial life. Fontenelle's sixth dialogue, which was added in 1687, neatly summarises the five main themes of his book:

> [1] the similarities of the planets to the Earth which is inhabited; [2] the impossibility of imagining any other use for which they were made; [3] the fecundity and magnificence of nature; [4] the consideration she seems to show for the needs of their inhabitants as having given moons to planets distant from the Sun, and more moons to those more remote; and [5] that which is very important – all that which can be said on one side and nothing on the other.[21]

Fontenelle's book was based on Lucretian logic. There must be *some* order to the cosmos, else how would the congruent and inquisitive mind be able to interpret it? For Fontenelle, the principle of plenitude was the organising hypothesis. With regard to the nature of plural worlds and their inhabitants, they were organised according to the 'infinite diversity that Nature placed in her works'.[22] The problem that this infinite diversity poses for the imagination is one of the main themes of Fontenelle's book. As one of his chief characters, the Marquise, explains, 'My reason is pretty well convinced, ... but my imagination's overwhelmed by the infinite multitude of inhabitants on all these planets, and perplexed by the diversity one must establish among them.'[23] But Fontenelle also offers a solution:

> It is not up to the imagination to attempt to picture all that. ... It is not proper for the imagination to go any farther than the eyes can. One may only perceive by a kind of universal vision, the diversity which Nature must have placed among all the worlds. All faces in general are made on

> the same model, but those of two large societies – European, if you like, and African – seem to have been made on two specific models, and one could go on to find the model for each family. What secret must Nature have possessed to vary in so many ways so simple a thing as a face? In the universe we're no more than one little family whose faces resemble one another; on another planet is another family whose faces have another cast. We can suppose the differences increase according to the distance one travels.[24]

Armed with the imagination, and using Atomism as his guide, Fontenelle tried to establish a link between terrestrial and extraterrestrial life. Just as today's speculative scholar might do the same with the principles of evolution, Fontenelle assumed Nature's diversity was not total. There may well be a modicum of fundamental comparability between inhabitants of other worlds and humans on Earth. In terms of the mechanism of this comparability, Fontenelle was understandably not specific. And yet his 'universal vision' evidenced the diversity of facial features on Earth, forced into conceivable bounds. Might not something similar also be at play in the cosmos at large?

This fusion of Copernicanism and Atomism also boasted 'economy of means' as one of its key characters. Witness one of the conversations between the narrator and the Marquise on the beautiful simplicity of the Copernican system. After informing the Marquise about the confusing complexities of the fated geocentric system, Fontenelle turns to the Copernican:

> 'I'm now going to propose a different one to you which satisfies all ... for it's one of a charming simplicity, which alone would make it preferable.' 'It would seem,' the Marquise interrupted, 'that your philosophy is a kind of auction, where those who offer to do these things at the least expense triumph over the others.' 'It's true,' I replied, 'and it's only by that means that one can catch the plan on which Nature has made her works. She's extraordinarily frugal. Anything that she can do in a way which will cost a little less, even the least bit less, be sure she'll only do it that way. This frugality, nevertheless, is quite in accord with an astonishing magnificence, which shines in all she does. The magnificence is in the design, and the frugality in the execution.'[25]

The imagining of alien life is central to Fontenelle's book. It begins with the most straightforward of references. Fontenelle communicates how the imagination plays an important role, for the narrator and his audience. For example, the narrator speaks of 'a certain young girl that has been spotted on the Moon through the telescope some forty years ago who has aged considerably',[26] and he adds, 'since everyone projects onto an object those ideas with which he is filled. Our Astronomers see on the Moon the faces of young girls, and it is quite possible that if women observed [the Moon], they'd see the handsome faces of the men there. As for myself, Madame, I think that I would see you there.'[27]

Life and ETI in the Renaissance Solar System

Fontenelle links the theory of pluralism to the Cartesian cosmology of vortices, the idea that planets orbit their respective suns in a vortex of celestial matter. He describes our own Solar System as a vortex around the Sun, and calls the Earth an imposter for previously assuming the Sun's position at the centre of the vortex, forcing the Sun and other planets to circle around her. Enter Copernicus, God-like heliocentric saviour, in Fontenelle's powerful history of corrective cosmology:

> Copernicus... lays violent hands on the different circles and solid spheres which were imagined by Antiquity. He destroys the first and breaks the others in pieces. Seized by a noble astronomical fury, he plucks up the Earth and sends her far from the centre of the universe, where she was placed, and puts the Sun in the centre, to whom the honour rightly belongs. The planets no longer turn around the Earth and enclose her in the circles they describe. If they light us, it's more or less by chance as we meet them in their paths. Everything turns around the Sun now, including the Earth, and as punishment for the long rest she was given, Copernicus charges her as much as he can with the same movements she had attributed to the planets and heavens. At last the only thing left of all this celestial train which used to accompany and surround this little Earth is the Moon that turns around her still.[28]

Just as the Sun has a family of planets, each planet has its family of inhabitants. During the course of *Conversations*, Fontenelle describes how the bodies of the Solar System, from the Moon to beyond, are inhabited. His vision of inhabited Venus is of a population that resembles 'the Moors of Granada, who were a little black People, scorch'd with the Sun, witty, full of Fire, very Amorous, much inclin'd to Musick & Poetry'.[29]

Using this curious form of adaptation, Fontenelle peoples the other planets. The proximity of the Sun renders Mercurians, 'so full of Fire, that they are absolutely mad; I fancy, they have no memory at all, ... that they make no reflections, and what they do is by sudden starts, and perfect hap-hazard; in short, Mercury is the Bedlam of the universe.'[30] Although Fontenelle maintained,

'If it is a certain truth, that Nature never gives life to any Creature, but where that Creature may live'[31], he was strangely scathing about Martians, 'Mars hath nothing curious that I know of . . . he is a little less than the Earth, . . . but let us leave Mars, he is not worth our stay.'[32]

Of Jupiter and Jovians, however, Fontenelle had much to say. Back in 1610, perhaps the greatest discovery Galileo had made with the spyglass was the essentially four new worlds in the Jupiter system. The four new planets were the first four moons of Jupiter, their significance to the case of Copernicanism Galileo had set out in *Starry Messenger*:

> Moreover, we have an excellent and exceedingly clear argument to put at rest the scruples of those who can tolerate the revolution of the planets about the Sun in the Copernican system, but are so disturbed by the revolution of the single Moon around the Earth while both of them describe an annual orbit round the Sun, that they consider this theory of the universe to be impossible.[33]

The moons in orbit around Jupiter also exposed the ancient fallacy that the Earth was the sole centre around which all revolved.

Fontenelle imagined an ETI on Jupiter. First referring to Galileo's revolutionary discovery of the Jupiter's moons, Fontenelle imagines how Jovian astronomers discover the Earth, 'after they have made the most curious Telescopes, and taken the clearest Night for their observations, they may have discover'd a little planet in the Heavens, which they never saw before'.[34] One Jovian astronomer believes Earth to be inhabited, but his colleagues laugh at him, 'if they publish their discovery, most People know not what they mean, or laugh at 'em for Fools'.[35]

The telescope and Renaissance solar systems

And so to the stars.

On the fifth evening of *Conversations* the Marquise wishes to know whether the stars too are inhabited. If the imagining is beyond our abilities to posit otherness, science is Fontenelle's measure. The narrative details further 'universal visions', which he calls upon to make flesh what is beyond imagining. Once more, Fontenelle uses his fusion of Atomism with the Cartesian theory of vortices.

The stars are judged by the narrator to be at least fifty million miles from the Earth, though an angered astronomer may put them even further out. Fontenelle explains that all the stars are suns and that each has its own vortex, many of which may harbour inhabited planets,

> 'May not the Worlds,' reply'd the Countess, 'notwithstanding this great resemblance between 'em, differ in a thousand other things; for tho' they may be alike in one particular, they may differ infinitely in others.' 'It is certainly true,' said I; 'but the difficulty is to know wherein they differ. One Vortex hath many Planets that turn round about its Sun; another Vortex hath but a few: In one Vortex, there are inferior or less Planets, which turn about those that are greater; in another perhaps, there are no inferior Planets; here, all the Planets are got round about their Sun, in form of a little Squadron; beyond which, is a great void space, which reacheth to the neighbouring Vortex's: In another place, the Planets take their course towards the out side of their Vortex, and leave the middle void. There may be Vortex's also quite void, without any Planets at all; others may have their Sun not exactly in their Centre; and that Sun may so move, as to carry its Planets along with it: others may have Planets which in regard of their Sun, ascend, and descend, according to the change of their Equilibration, which keeps them suspended. But I think I have said enough for a Man that was never out of his own Vortex.'[36]

There is little doubt that the invention of the telescope was instrumental to the spread of Copernican pluralism, and with it a re-birth of Atomism.

The technology that made up the spyglass had long been known. The history of its optics, however, is a captivating example of the social relationship between theory and technique. Even ancient civilisations had known of the curious lens effect made by, say, transparent crystals, or glass spheres full of water. The magnifying effects were a fascination. But those observations came too soon. In the texts of ancient Greek and particularly medieval Arab scholars, it is clear that a theory of optics existed. The technology to magnify had also existed. And clearly the potential demand for the spyglass in commerce and warfare was obvious. Theory was never married to practice, however. Texts were never tallied up to devices.

With regard to its influence on the development of pluralism, the curious evolution of the telescope can be explained in two connected ways. First, a social explanation: it is very unusual in history for scholars and craftsmen to work together, given their difference in social class. Such was the case throughout the pluralist dark ages up until the European Renaissance, one of the distinguishing features of which was the intensity and distribution of the collaboration between the work of the mind and the work of the hand.

Second, it is worth considering the old proverb 'a little knowledge is a dangerous thing'. Ancient observations made through transparent crystals, or glass spheres, do not merely show things bigger, and better. They also deceive.

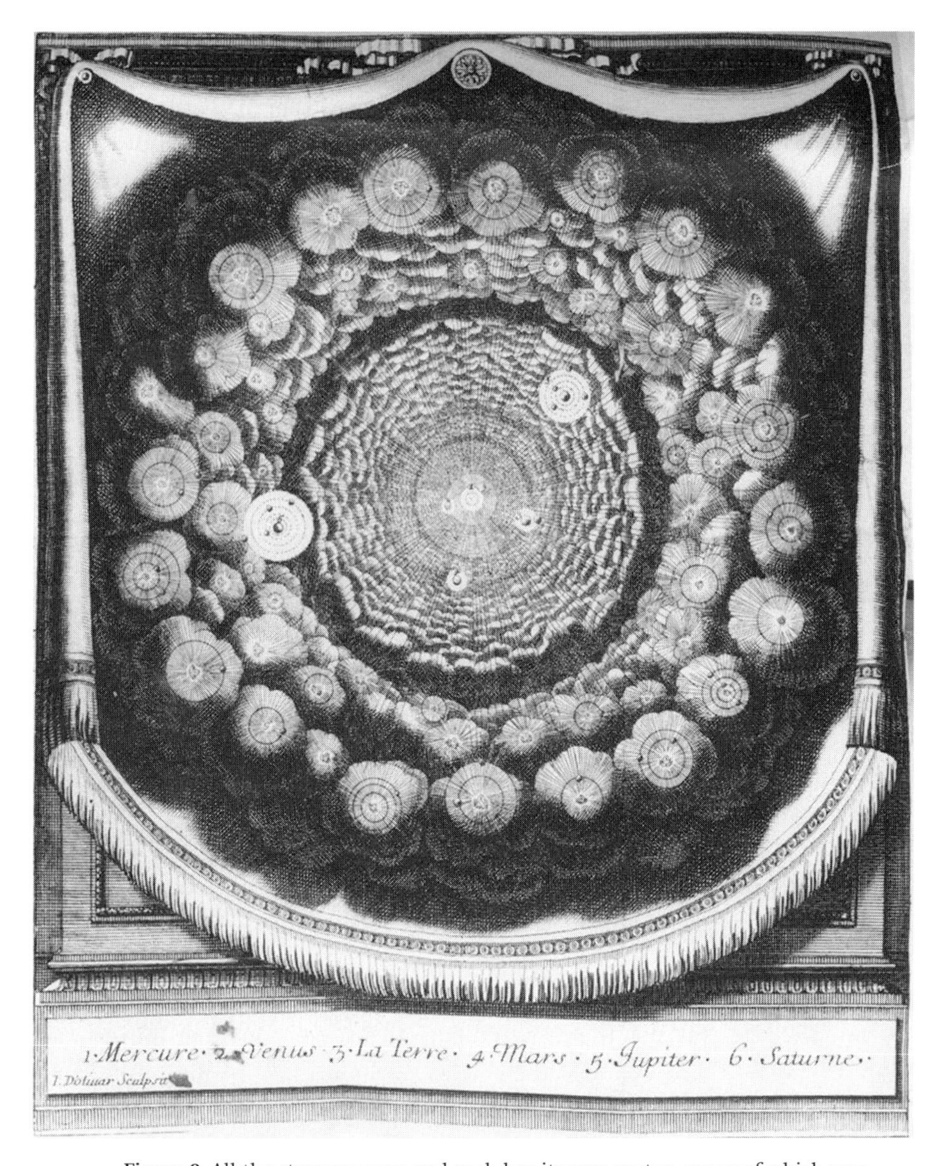

Figure 9 All the stars are suns and each has its own vortex, many of which may harbour inhabited planets. Bernard le Bovier de Fontenelle, *Entretiens sur la Pluralite des Mondes*, 1686.

To the superstitious mind, a mind not married to rational theory, optical illusions can make one wary. Sight is the most paradoxical of senses. It is at once the most reliable, and yet the most deluding. Without theory, maybe you cannot really trust what you see.

Eventually, Islamic scholars Ibn Sahl and Ibn al-Haytham had, in the tenth and eleventh centuries, refined the technical knowledge necessary for the production of spectacle lenses. But the earliest known working 'scopes were lens-based refracting telescopes that materialised in the Netherlands in 1608. Legend has it that a couple of children were playing with lenses in the shop of one Hans Lippershey. The kids gazed through one lens at another in the shop window, around the year 1600. Curiously, the blend of lenses made things outside the shop seem closer.

The very fact that no scientist was needed to invent the telescope is worthy of note. The device was long overdue, and the means of making a spyglass had probably been present for at least three hundred years. Only with the sheer volume of lens manufacture through the increased wealth of the sixteenth century was the discovery brought about by chance. But it made a world of difference to pluralism.

The introduction of the telescope greatly helped promote the concept of plural worlds. Galileo's use of the spyglass is a case in point. He discovered the moons of Jupiter, the phases of Venus, the terrestrial nature of the Moon, and thousands upon thousands of stars never picked out before. Galileo's findings provided immense impetus to the Copernican viewpoint on the cosmos, and also boosted Atomist speculation. Maybe this universe is, as the Atomists said, a cosmos vast in scale, perhaps even infinite, and abundantly seeded with life throughout.

Atomist ideas also helped with imagining alien life. While the evidence gifted by the spyglass greatly encouraged speculation on life beyond the Earth, it was insufficient to further the discussion beyond mere speculation. Those brave enough to speculate and use their imagination turned to Atomism for their worldview. The new and imagined cosmos was in constant motion, a universe of atoms and void, a cosmos of innumerable suns and innumerable worlds, some devoid of life, but many others replete not just with life, but intelligent life.

The telescope was not the only weapon of discovery used in the Atomist pluralist cause. There was also the microscope. Once again, the Netherlands was at the forefront in the development of optical instruments. Dutch eyeglass makers Hans Janssen and Hans Lippershey (he of the early telescope) were among the early developers credited, round about 1590. The name 'microscope' is credited to German botanist Giovanni Faber, who coined the name for fellow Lincean[37] Galileo's compound microscope in 1625.[38]

The microscope unveiled evidence of previously unknown, though minuscule, life. And yet the implication was clear. Beyond our perception lay remote worlds, realms that may endlessly reveal themselves as lens-grinding refined, and as science marched forward. These small and remote worlds too were put to the pluralist cause. Might not the cosmos at large also harbour unseen worlds?

Both Fontenelle and Cyrano used the microscope in their case for remote and hidden worlds. In *Conversations*, after declaring his disbelief for a unique inhabited Earth, the narrator explains to the Marquise that, of the creatures inhabiting the Earth, those evident to the naked eye are nowhere near all that exist: 'We see from the elephant down to the mite; there our sight ends. But beyond the mite an infinite multitude of animals begins for which the mite is an elephant, and which can't be perceived with ordinary eyesight.'[39] So, the narrator implies, the system of the world, as humans experience it, may also be projected in the opposite and cosmic direction.

Cyrano's *Lune* had engaged the reader in a more imaginative account of the microscopic world. Writing a few decades before Fontenelle, Cyrano used a microscopic structure, contained in a larger one, the human body, and in so doing transformed the body into a world. Fontenelle must have been familiar with Cyrano's account, and both authors elegantly combine the microscopic and cosmic perspectives. In Cyrano's case, the comparison comes when Dyrcona and a philosopher of the Moon talk of how the cosmos consists of innumerable worlds of different scales, locked into each other. The world within Dyrcona's body is described:

> For do you find it hard to believe that a louse should take your body for a world, or that, when one of them travels from one of your ears to the other, his friends should say that he has voyaged to the ends of the earth, or that he has journeyed from pole to pole? Why, doubtless this tiny people take your hair for forests of their country, your pores full of sweat for springs, your pimples for lakes and ponds, your abscesses for seas, your streaming nose for a flood; and when you comb your hair backwards and forwards they think this is the ebb and flow of the ocean tides.[40]

And so we have Cyrano's system of the world. In this ingenious fusion of microscopic and cosmic perspectives, Cyrano likens the movements that the human form creates on its tiny inhabitants to the prospect of the revolutions of heavenly bodies upon humans. If, contained within the human form, the microscope unveils an entire world, what better vision could there be for the idea of a cosmic plurality of inhabited worlds. Each piece of creation is at once part and whole, varying with viewpoint. And on it continues, Cyrano's system of the cosmos is presented as a continuum of smaller worlds, nested into greater ones, an onion-layered universe.

Cosmotheoros

But soon, the Moon lost its lustre.

The quality of the spyglass improved. And as the refinement in lens grinding and telescope technology progressed, a more detailed view of the cosmos was afforded. Through the 'scope, astronomers discovered that the craters in the dark parts of the lunar surface, the 'old spots' known to the naked eye, could not be covered in water, as had been thought. On top of these doubts about the existence of lunar water, it began to dawn that the Moon was unlikely to have an atmosphere. Without water and air, it became harder to imagine lunar creatures being anything like life on Earth. And, with the Moon lost, one of the main arguments in favour of pluralism was also lost.

The loss of the Moon was a problem too for Christiaan Huygens. Considered as eminent as Newton among the leading physicists of the second half of the seventeenth century, Huygens was among the most accomplished astronomers of the period. His early telescopic work had clarified the nature of Saturn's rings and discovered the moon Titan. Indeed, Huygens' labour in the field of physics led also to the invention of the pendulum clock, the wave nature of light, and the discovery of centrifugal force. He is also considered the first theoretical physicist to use formulae in physics.

It is arguable that neither before nor since has a scientist of the stature of Huygens devoted an entire book to the question of extraterrestrial life. His thoughts were set out, prior to his death in 1695, in a volume called *Cosmotheoros*. Like Newton, Huygens had a religious calling. He was Protestant, but with leanings toward scepticism. His treatise on extraterrestrials was published posthumously in 1698 in Latin, and under the full title *Κοσμοθεωροσ, sive de terris coelestibus earumque ornatu conjecturæ*. The book was translated and published in English in the same year, under the title, *The Celestial Worlds Discover'd: or, Conjectures Concerning the Inhabitants, Plants and Productions of the Worlds in the Planets*. French, German, Russian, and of course Dutch translations also followed.

Huygens' book attracted much attention; not the least of which was the approval of John Flamsteed, the director of the Greenwich Observatory. Flamsteed approved of *Cosmotheoros* so much he recommended the book to the vicar of Greenwich, Archdeacon Thomas Plume. And Plume's fascination with Huygens' work was lucrative indeed. He bequeathed £1902 to Cambridge University to 'erect an observatory and to maintain a professor of astronomy and experimental philosophy, and to buy or build a house with or near the same'.[41] And so was established the Plumian chair of Astronomy at Cambridge, a charming example of the way in which the alien has influenced mainstream physics.[42]

For Huygens, water was life's matrix. He thought that life on other worlds was similar to that on Earth. His reasoning was this: the availability of liquid water was vital for the existence of life. Even though the properties of water should vary from planet to planet, it still acted as life's medium on the different worlds. Huygens was aware of the fact that the kind of water found on Earth would instantly freeze on Jupiter, and vaporise on Venus. He observed dark and bright spots on the surface of some of the planets, such as Mars and Jupiter, and suggested the spots could only be justified by the existence of water and ice on those worlds:

> That the Planets are not without Water, is made not improbable by the late Observations: For about Jupiter are observ'd some spots of a darker hue than the rest of his Body, which by their continual change show themselves to be Clouds ... Mars too is found not to be without his dark spots... but whether he has Clouds or no, we have not had the same opportunity of observing as in Jupiter ... Since 'tis certain that Earth and Jupiter have their Water and Clouds, there is no reason why the other Planets should be without them. I can't say that they are exactly of the same nature with our Water; but that they should be liquid their use requires, as their beauty does that they should be clear. For this Water of ours, in Jupiter or Saturn, would be frozen up instantly by reason of the vast distance of the Sun. Every Planet therefore must have its Waters of such a temper, as to be proportion'd to its heat.[43]

Huygens was unrelenting in his belief of aliens. He wanted to populate every nook and cranny of the cosmos. His main argument was this: there was a degree of uniformity to the different worlds in the cosmos, based on water. In an approach he shared with Fontenelle, Huygens also believed that the similarities between the Earth and other worlds grew weaker with distance. Those worlds close at hand, it follows, should be strikingly similar to our home planet. But the Moon was a problem. As the Moon was the Earth's closest neighbour in a vast and Atomist universe, the lunar world's lack of water and atmosphere would have invalidated his pluralist hypothesis.

Huygens hatched an escape route. To save the notion of the habitability of the planets, he argued that the negative evidence of the Moon did not serve as symbolic for the real planets. Establishing a correspondence between the Moon and planets and social class, Huygens argued that the planets and moons of the Solar System should be evaluated in terms of their very different rank on the cosmic scale:

> And here one would think that when the Moon is so near to us, and by means of a Telescope may be so nicely and exactly observ'd; it should afford us matter for more probable Conjectures than any of the other remote Planets. But it is quite otherwise, and I can scarce find anything to say of it, because I have not a Planet of the same nature before my eyes, as in all the primary ones I have. For they are of the same kind with our Earth; and seeing all the Actions, and every thing that is here, we may make a reasonable Conjecture at what we cannot see in those Worlds. But this we may venture to say, without fear, that all the

> Attendants of Jupiter and Saturn are of the same nature with our Moon, as going round them, and being carry'd with them round the Sun just as the Moon is with the Earth. Their Likeness reaches to other things too, as you'll see by and by. Therefore whatsoever we can with reason affirm or fancy of our Moon (and we may say a little of it) must be suppos'd with very little alteration to belong to the Guards of Jupiter and Saturn, as having no reason to be at all inferior to that.[44]

As if by default, Huygens argues for ETI on other worlds. If the Moon is not inhabited, as the observations suggest based on the absence of water and an atmosphere, this says little about life on other worlds, save for those of similar rank, i.e. other moons. Which means that other planets, especially those superior and majestic worlds, such as Saturn and Jupiter, should be equal in all ways to the Earth, intelligent inhabitants included.

So Huygens imagined a plural cosmos as a reflection of bourgeois European society. By transplanting a hierarchical social structure onto planets and moons, he was able to underplay the impact of the Moon's negative evidence. By spinning the idea of the cosmos as a class society, he was able to leave intact the notion of inhabited worlds.

When Galileo's discovery of the Jovian moons was confirmed in Kepler's short pamphlet *Observation-Report on Jupiter's Four Wandering Satellites*, it was the first independent confirmation of the Jupiter system. It was also history's first use of the word 'satellite', to mean a moon in orbit around a celestial body. Perhaps the mystical Kepler wished to preserve Jupiter's godlike cachet. In Roman mythology, Jupiter was king of the gods. And as such, powerful men were usually orbited by a knuckle of bodyguards, or *satellitem* in Latin. Rather fitting, then, that Kepler should use satellites for the Galilean moons in seemingly protective orbit about the planet of the same name.

Huygens keeps Jupiter's godlike cachet intact. If the 'guards' and 'satellites' of the mighty Jupiter do not show signs of life, there is little need to worry. It is totally consistent with Huygens' universal vision. The role of the satellites is not to show signs of life, but to serve their master. And so the idea of social hierarchy helps to imagine a system of the world in which evidence is beginning to play a part.

Next Huygens contemplates colonisation. He rejects Kepler's idea that the Moon is inhabited. Huygens also discards Kepler's idea of the lunar craters as some kind of fortress, built by Moon folk. He suggests that the crater phenomena are far too big to have been built, and puts their presence down to natural causes. It is in this context that he now considers planting colonies on the moons of the Solar System. Such imperial ambition is aimed at moons, and not planets.

Indeed, no mention is made of the possibility of colonising other planets. Once more, the distinction of social class comes into play. Worlds whose owners are thought of as social equals must be respected. The moons, however, are available for piracy and plunder.

God among the aliens

The pluralism of Huygens was godly in nature.

His principle of plenitude was teleological in that he supposed nothing had been made without purpose. Teleology is the doctrine of design and purpose in the material world. So, the planets that spun about the Sun were not wasteful in their creation, but must offer habitation to various forms of life. As Huygens declares, 'For all this Furniture and Beauty the Planets are stock'd with seem to have been made in vain, without any design or end, unless there were some in them that might at the same time enjoy the Fruits, and adore the wise Creator in them.'[45] Since there must be countless stars beyond our ken, they cannot have been created for us. They must be a source of light and life to other worlds. And since the Christians believed God's Creation finds fulfilment in the contemplation of rational beings, it made more religious sense, depending upon your convictions of course, that the deep cosmos was inhabited by ETI.

Cosmotheoros is replete with such arguments. In answer to those who suggest the universe was created solely for man, Huygens writes,

> And these Men themselves can't but know in what sense it is that all things are said to be made for the use of Man, not certainly for us to stare or peep through a Telescope at; for that's little better than nonsense. Since then the greatest part of God's Creation, that innumerable multitude of Stars, is plac'd out of the reach of any man's Eye; and many of them, it's likely, of the best Glasses, so that they don't seem to belong to us.[46]

And so Huygens concluded the multitude of stars invisible to man would nonetheless be seen by aliens on other worlds. Another aspect to this godly approach was the teleological idea of economy. This was the notion that nature realises its potential with the least possible means, and, should the opportunity arise, different ends by a single means. Such economy of means was claimed by the more progressive teleologists to be one of the main advantages of the Copernican system, over the old geocentric cosmos. In the words of Huygens, 'This is the now commonly receiv'd System, invented by Copernicus, and very agreeable to the frugal Simplicity Nature shows in all her Works.'[47]

As to what form these off-Earth creatures took, Huygens believed they would be like us. He validates his argument by saying that the way things are on Earth is so perfect that there is little room for variation, since Nature cannot have created the other worlds as anything less than perfect:

> The Stature and Shape of Men here does show forth the Divine Providence so much in its being so fitly adapted to its design'd Uses, that it is not without reason that all the Philosophers have taken notice of it nor without probability that the Planetarians have their Eyes and Countenance upright, like us, for the more convenient and easy Contemplation and Observation of the Stars. And the Wisdom of the Creator is so observable, so praiseworthy in the position of the other Members; in the convenient situation of the Eyes, as Watches in the higher Region of the Body; in the removing of the more uncomly parts out of sight as 'twere; that we cannot but think he has almost observed the same Method in the Bodies of those remote Inhabitants. Nor does it follow from hence that they must be of the same shape with us. For there is such an infinite possible variety of Figures to be imagined, that both the Oeconomy of their whole Bodies, and every part of them, may be quite distinct and different from ours.[48]

Where Huygens led, Leibniz later followed. In 1710 in his work *Essais de Théodicée sur la bonté de Dieu, la liberté de l'homme et l'origine du mal* (*Essays on the Goodness of God, the Freedom of Man and the Origin of Evil*), German polymath Gottfried Wilhelm Leibniz was to declare our planet the best of all possible worlds. Huygens' argument, made a couple of decades before, is strikingly similar, but with respect to the peopling of other planets. The comparison is in the principle of necessity, which requires that, 'if a rational God created the world, its order had to agree with and be determined by a rational logic'.[49]

Cosmotheoros had been based on the principle of uniformity in nature. The idea goes that even while Nature may have created things with infinite variety on other worlds, 'such as neither our Understanding nor Imagination can conceive',[50] that does not mean that because this variety, 'may be Infinite, and out of our comprehension and reach, that therefore things in reality are so'.[51] It seems a rather paradoxical principle. Huygens uses it as a licence for imagining other worlds, with their animals, plants, and people, in all cases almost identical to those on Earth. And like Leibniz after him, he justifies such geocentric notions with the curious get-out clause: Earth is so perfect there is precious little room for variation, since nature, through God, cannot have created the other worlds too as anything less than perfect.

In his godly *Cosmotheoros*, Huygens is rather scathing about the ungodly Epicureans. True, he is also damning about Kepler, who wrote 'pretty Fairy Stories of the men in the Moon',[52] and Fontenelle, who failed to 'carry... the business any farther'[53] than Bruno. But his most stinging criticism is for the materialist Epicureans and Cartesians, who conjure up living creatures as 'haply jumbled together by a chance Motion of I don't know what little particles'.[54]

Then there was Newton.

After the resurrection of Atomism, and the labours of workers such as Copernicus, Galileo, and Bruno, the Christian picture of creation was fading. The influence of world-famous scholars, such as Galileo and his influential allies, were held in check by the power of the Church. When Galileo had supported the new astronomy, he had been warned in 1616 by a Holy Office that wished to maintain its longstanding religious and political influence in Europe. As Thomas Kuhn so powerfully put it:

> When it was taken seriously, Copernicus' proposal raised many gigantic problems for the believing Christian. If, for example, the Earth were merely one of six planets, how were the stories of the Fall and of the Salvation, with their immense bearing on Christian life, to be preserved? If there were other bodies essentially like the Earth, God's goodness would surely necessitate that they, too, be inhabited. But if there were men on other planets, how could they be descendents of Adam and Eve, and how could they have inherited the original sin, which explains man's otherwise incomprehensible travail on an Earth made for him by a good and omnipotent deity? Again, how could men on other planets know of the Saviour who opened to them the possibility of eternal life? Or, if the Earth is a planet and therefore a celestial body located away from the centre of the universe, what becomes of man's intermediate but focal position between the devils and the angels? If the Earth, as a planet, participates in the nature of celestial bodies, it can not be a sink of iniquity from which man will long to escape to the divine purity of the heavens. Nor can the heavens be a suitable abode for God if they participate in the evils and imperfection so clearly visible on a planetary Earth. Worst of all, if the universe is infinite, as many later Copernicans thought, where can God's Throne be located? In an infinite universe, how is man to find God or God man?[55]

Tricky. It is quite apparent why most thinkers chose not to be as bold as Huygens. Descartes, for instance, was always very cautious when it came to pluralism. After noting that Christ's blood had redeemed many men, Descartes went on,

> I do not see at all that the mystery of the Incarnation, and all the other advantages that God has brought forth for man, obstruct him from having brought forth an infinity of other very great advantages for an infinity of other creatures. And although I do not at all infer from this that there would be intelligent creatures in the stars or elsewhere, I also do not see that there would be any reason by which to prove that there were not; but I always leave undecided questions of this kind rather than denying or affirming anything.[56]

Even so, Descartes did a great service for pluralism. The Cartesian system of the world was the most important pluralist cosmology of the seventeenth century, one in which each star was a sun surrounded by satellite planets. Descartes gifted an open door to pluralism through which only the brave would follow.

Isaac Newton was not among them.

As arguably one of the most accomplished scientists of the Scientific Revolution, his contemporaries would carefully scrutinise his texts, as they became published. Those looking for wisdom or word on life beyond the Earth would have been disappointed. Like Descartes, Newton was reluctant to commit, and was often openly hostile to Epicurean cosmogony.

Newton's two most important works were *Principia*,[57] the text that was to dominate the scientific view of the physical universe for the next three centuries, published in 1687, and *Opticks*, published in 1704. Both texts contained profoundly relevant findings for pluralism. In his system of the world in *Principia*, Newton set out his theory of universal gravitation, which had a number of implications for life elsewhere in the cosmos. For one thing, his calculation of planetary masses in proposition 8, book 3, suggested that Earth was six times as dense as Saturn. If correct, it raised serious questions as to whether Saturn had a solid enough surface to support its assumed inhabitants. Such is the influence of number.

His case becomes even more compelling. When the 1713 edition of *Principia* was published, it contained Newton's 'General Scholium', a discourse on various topics, but including a best means fit of his system of the world to theology. The two are naturally incompatible, but Newton was not to be deterred:

> This most elegant arrangement of the sun, planets, and comets could not have arisen but by the plan and rule of an intelligent and powerful being. And if the fixed stars be centres of similar systems, all these, constructed by a similar plan, will be under the rule of One, especially because the light of the fixed stars is of the same nature as the light of the sun, and all systems send light into all mutually. And so that the

> systems of the fixed stars should not fall into each other mutually, he will have placed this same immense distance among them.[58]

The last sentence is typical of the way in which Newton evaded the alien. His cosmology has clear implications for extraterrestrial life, and yet he is cautious about such religiously controversial conclusions. It was the same in a query he added to his *Opticks* in 1706. If there was such a thing as an atomic theory of matter, Newton was not going to allow any Epicurean and atheistic conclusions to follow:

> And since Space is divisible in infinitum, and Matter is not necessarily in all places, it may be also allow'd that God is able to create Particles of Matter of several Sizes and Figures, and in several Proportions to Space, and perhaps of different Densities and Forces, and thereby to vary the Laws of Nature, and make Worlds of several sorts in several Parts of the Universe. At least I see nothing of Contradictions in all this.[59]

According to Newton, God did it. Everything. In the coming centuries, Newtonianism became enormously influential. It was an intellectual agenda, one that applied Newtonian principles to many avenues of enquiry. And not just in physics and mathematics. Philosophy, politics, and theology would also come under its considerable sway.

The Newtonians

The man who in 1702 succeeded Newton as Lucasian Professor of Mathematics at Cambridge was a pluralist. William Whiston had already promoted life on other planets in his 1696 book *New Theory of the Earth*. Twenty years later, in his *Astronomical Principles of Religion*, Whiston went one better by proposing inhabitants of the interiors of the Earth, Sun, planets, and comets. He went even further. The planetary atmospheres too were complete with 'not wholly Incorporeal, but Invisible Beings'.[60]

Between those two books, Whiston had been an advocate of the Newtonian system. The publication of his Cambridge astronomy lectures had incorporated the idea of stars as suns, but in 1710 Whiston had lost the Lucasian chair due to religious bigotry. Charges of religious heterodoxy, specifically his alleged Arianism, had provided Whiston with the opportunity to take pluralism on the road, lecturing at many venues in order to spread the extraterrestrial word.

Even the most modest lecture and location can have remarkable results. One of Whiston's talks was at the Button Coffeehouse in London in 1715. Here, Alexander Pope was introduced to the new pluralist universe, which he recounted later in his *Essay on Man*. Pope's essay declared that no proper account of man should be without due consideration of the extraterrestrial plane, which English astronomer Thomas Wright and German philosopher Immanuel Kant had duly quoted in their own accounts of the cosmos:

> He, who thro' vast immensity can pierce,
> See worlds on worlds compose one universe,
> Observe how system into systems runs,
> What other planets circle other suns,
> What vary'd Being peoples ev'ry star,
> May tell why heav'n has made us as we are.[61]

Whereas, a century before, John Donne had envisioned a vast pluralist universe of cosmic chaos, Pope imagined a Newtonian system of order, again quoted by Kant:

> Who sees with equal eye, as God of all,
> A hero perish, or a sparrow fall,
> Atoms or system into ruin hurl'd,
> And now a bubble burst, and now a world.
> Superior beings, when of late they saw
> A mortal man unfold all Nature's law,
> Admir'd such wisdom in an earthly shape,
> And shew'd a Newton as we shew an Ape.

Another important acolyte to the Newtonian cause was Anglican scholar Richard Bentley, soon to be Master of Trinity College, Cambridge. Bentley was a prominent academic in his own right, known for his literary and textual criticism; he was the first Englishman to be ranked with the great heroes of classical learning, inspiring generations of subsequent scholars. Bentley was a young man of thirty when he was invited by his contemporaries to inaugurate in 1692 a series of lectures, funded by the famous chemist Robert Boyle, devoted to 'proving the Christian religion'.[62] A tall order, so why not call in another 'expert'. Before completion of the last two of the eight lectures he delivered, Bentley wrote to Newton, asking whether his system of the world, presented in *Principia*, was supportive of religion.

Newton was delighted. He framed his response to Bentley in the form of four letters, which were themselves published in 1756, three decades after Newton's death. The letters cover various topics, including Bentley's reference to the idea of a succession of worlds. A number of Church fathers, including Augustine and Nicole Oresme, a French philosopher of the later Middle Ages, had interpreted Aristotle on this topic. The idea was one of a plurality of worlds in time, rather than space.

The unimaginative Newton was having none of it. For, 'the Growth of new Systems out of old ones, without the Mediation of a divine Power, seems to me apparently absurd.'[63] There are none so blind as those who will not see. Newton also discarded any materialist accounts of the cosmos. For example, he refuted any purely mechanical explanation of the creation of the Solar System, on the grounds that the 'shining' matter of the Sun could not have been separated from the 'opaque' matter of the planets.[64] Such was Newton's hostility, not only to Epicurean cosmogony, but also to any materialist and evolutionary approach to the cosmos.

Bentley was also hostile to ungodly Epicureanism. Indeed, he confronted a number of Epicurean themes in his seventh and eighth lectures. In his seventh lecture he made it clear that he endorsed pluralism, 'because every Fixt star is supposed by Astronomers to be of the same Nature with our Sun; and each may very possibly have Planets about them, though by reason of their vast distance they be invisible to Us: we will assume this reasonable supposition.'[65] In his eighth lecture, Bentley discourses on other worlds, and the context in which he deals with the idea of extraterrestrial life is a compelling one.

For Bentley was faced with a problem. His challenge was to prove God's beneficent design evident in nature, not in the limited walled-in cosmos of the ancients, but in the vast universe of the late seventeenth century, within a world of the Copernican revolution and Newtonian mechanics. What makes Bentley's case so compelling is this: he was the first author to put the pluralist case from within the godly framework of the system that Newton had created, five years earlier, in his *Principia*.

Geocentrism died hard. Proceeding with caution, and suspecting a sticky wicket, Bentley suggests God could indeed have created the entire universe for man as 'the Soul of one vertuous and religious Man is of greater worth and excellency than the Sun and his Planets and all the Starrs in the World'.[66] But Bentley also envisioned plural worlds, as the stars could not be explained as existing for themselves, clumps of mere matter deprived of people, and he 'dare' not believe that the stars provide an advantage to man. So, the heavens 'were formed for the sake of Intelligent Minds', and just as the Earth was made for man, 'why may not all other Planets be created ... for their own Inhabitants which have Life and Understanding?'.[67]

And yet we begin to glimpse the true alien. One of the questions Bentley had to confront, as was indicated in the above quote from Thomas Kuhn, was the relation of extraterrestrials to the fall of Adam and the incarnation of Christ, both momentous events in the drama of the Christian narrative. Bentley is shrewd. He rejects the problem on the basis that extraterrestrials need not be men. Indeed, he imagines that God 'may have made innumerable Orders and

Classes of Rational minds; some higher in natural perfections, others inferior to Human Souls. [These] would constitute a different Species.'[68]

God had a hand in the Goldilocks Zone too. According to Bentley in his eighth lecture, our planet was placed by Him at such a distance from the Sun, known today as the Goldilocks Zone, that we are neither too cold nor too hot. But such pious thoughts often produce other problems; in this case, how does God ensure, for all must surely be right with His plan, that extraterrestrials are to survive on Mercury and Saturn, placed by Him so differently from the Sun? Bentley had an answer:

> The Matter of each Planet may have a different density and texture and form, which will dispose and qualifie it to be acted on by greater and less degrees of Heat according to their several Situations; and that the Laws of Vegetation and Life and Sustenance and Propagation are the arbitrary pleasure of God, and may vary in all Planets . . . in manners incomprehensible to our Imaginations.[69]

What Newton made of Bentley's radical excursions into the alien is a captivating question, but one which, unfortunately, must be left to speculation. Bentley's is a curious cosmos, and one quite different from the 'universal laws' approach of the Atomists. To enable his godly logic, Bentley had to conjure up not only other worlds, but also worlds made of matter with different dispositions and integrity and lifeforms, all governed by different laws set by a seemingly whimsical Deity. It was a confusing cosmos indeed.

And yet this quaint concoction was soon to become commonplace. The heady mix of Newtonianism, Christianity, and pluralism was to be a popular combination in the decades that followed Newton and Bentley. With its tendency to oppose Epicurean Atomism, with its evolutionary approach to the cosmos, the new concoction signalled something of a reaction. Another proponent of the black art was Reverend William Derham. In 1714, while chaplain to the future King George II, Derham published *Astro-Theology, or A Demonstration of the Being and Attributes of God from a Survey of the Heavens*. The book was written as a sister volume to his *Physico-Theology*, and its main aim was to do for the extraterrestrial what its companion volume had done for the terrestrial.

Both of Derham's books proved very popular. By 1777, *Astro-Theology* had gone through fourteen English and six German editions. The book struck an easy balance in the vein of the new concoction. Speculations on God's design of deep space appealed to lay readers, to whom it taught plenty of astronomy, while the initiated were treated to more in-depth research. In Germany in particular, Derham's *Astro-Theology* spread the Newtonian word, promoting a system of the world from which it set its compass. The book's fabric is rich in pluralist thought, indebted also to Huygen's *Cosmotheoros*, though Derham begged to differ with the Dutchman on the question of life and water on the Moon.

Derham's *Astro-Theology* is a milestone in the history of pluralism for one other prominent reason: his depiction of world systems. In the book's preliminary discourse, Derham describes the historical systems of the world as 'Ptolemaick', which he discards, and 'Copernican', which he embraces, but only as a precursor to his 'New System' of Newtonianism: 'There are many other Systemes of Suns and Planets, besides that in which we have our residence; namely, that every Fixt Star is a sun and encompassed with a Systeme of Planets, both Primary and Secondary as well as ours.'[70] This new system of the world 'is far the most magnificent of any; and worthy of an infinite CREATOR'.[71]

And so Newton's system of the world essentially re-established the integrity of design, which had been shattered by the emergence of Atomism. The Christian picture of creation stayed more or less untouched, save some few concessions to pluralism, at least for the time being.

Man was still made in the image of God. But what of the word of Newton himself? There are another couple of references of note. The first dates from after 1710, and was first unearthed when David Brewster was writing his 1855 biography of Newton. Brewster claims that Newton believed in extraterrestrials, as this passage shows Newton stating that, at the final judgement, Christ

> will give up his kingdom to the Father, and carry the blessed to the place he is now preparing for them, and send the rest to other places suitable to their merits. For in God's house (which is the universe) are many mansions, and he governs them by agents which can pass through the heavens from one mansion to another. For if all places to which we have access are filled with living creatures, why should all these immense spaces of the heavens above the clouds be incapable of inhabitants?[72]

Newton seems to have been torn about the alien. In his original manuscript, Newton has crossed out that portion of the above passage that follows after 'many mansions'. In addition, he has added what seems to be a safer, replacement sentence, 'We are also to enter into societies by Baptism & laying on of hands & to commemorate the death of X in our assemblies by breaking of bread'.[73]

The final piece of evidence in the Newton jigsaw comes from an encounter that occurred two years before his death. In 1725, British Member of Parliament and Master of the Mint John Conduitt, the husband of Newton's niece, visited him. In the conversation that followed, Newton disclosed his feelings on pluralism, his thoughts about extraterrestrials, and about other questions of the cosmos. Conduitt's account of that meeting was published in 1806.

In conversation with Conduitt, Newton admitted to believing that 'revolutions' occur in the systems of stars, 'all which he took to be suns enlightening other planets, as our sun does ours'.[74] In Newton's view, these revolutions were the result of solar emissions, forming into satellites of planets:

> Vapours and light emitted by the sun, which had their sediment as water, and other matter had, gathered themselves by degrees, into a body, and attracted more matter from the planets; and at last made a secondary planet (viz. one of those that go round another planet), and then by gathering to them and attracting more matter, became a comet, which after certain revolutions, by coming nearer and nearer to the sun, had all its volatile parts condensed, and became a matter fit to recruit, and replenish the sun.[75]

This phenomenon, in Newton's view, was devastating. Though the emissions would replenish suns, the extreme heat created would destroy life on 'the earths served by those suns'.[76] According to Conduitt, 'He seemed to doubt whether there were not intelligent beings superior to us, who superintended these revolutions of the heavenly bodies, by the direction of the Supreme Being.'[77] As for the question of how such worlds may be inhabited after such revolutions, Newton offered, 'that required the power of a Creator'.[78]

So, Newton even seems to have believed in superior ETI. Assuming Conduitt's account to be accurate, we face the fact of a pluralist Newton, albeit a publicly reluctant one. He appears to have believed in planets in orbit about other suns, a succession of 'worlds' on this Earth, and almost certainly life at large in the cosmos. Though admittedly Newton's 'intelligent beings superior to us'[79] may be angels as much as aliens.

God's physicist

The very success of Newton's system carried with it a dialectic of disadvantage. Born in the same year that Galileo had died, Newton came from the new rural middle class of England that had already shaped Cromwell and the parliamentary officers. His physics and his fluxions marked the culmination of the labour of many generations of mathematicians, and finally brought the end of Aristotle's world-picture. A system operated on God's order had been replaced by Newton's mechanical system that operated simply by natural law, with no need of divine agency.

But Newton left a loophole. Unsure about his ungodly system, he allowed the prospect of divine intervention to maintain its stability. Having come under the influence of Platonists while at university, Newton had stopped short of any fundamental criticism of the existence of a divine plan. It was left to Laplace to close the loophole, and God's intervention was dispensed with. Newton, however, wished to ask no further questions, disguising his ignorance of origins by proposing the will of God at the beginning of creation.

By the time Newtonianism took shape, a new compromise was needed. The upheaval of the Renaissance and Reformation was over. Now, a system of concessions between religion and science was needed as much as a compromise between republic and monarchy, rising bourgeoisie and incumbent nobility. Newton's system of the world was a considerable concession on the part of the church. For in conceding the system, God's hand could no longer be seen in each and every cosmic or earthly event. God was relegated to the margins, to the details, to the creation and organisation of the universe, like some kind of cosmic accountant. God had, essentially, like his anointed ones on planet Earth, become a constitutional monarch.[80]

But science also had a price to pay. The other part of the unwritten bargain was a rejection of the more radical Atomism in favour of a vow not to trespass into the proper field of religion, and the world of man's life. Indeed, it was this compromise that was preached so formidably by Bentley in his Boyle sermons of 1692, and was to last until the so-called Darwinian revolution in the nineteenth century. The other price science paid was this: Newton's system seemed so perfect that it positively discouraged scientific advance for the next century. Newton's authority was more enduring than his system, and the whole tenor he gave to science was taken so much for granted that the acute limitations it implied were not fully known until Einstein.

The road to hell is paved with good intentions. Newton had meant his philosophy to be a merely mathematical expression, but the most immediate effect of his ideas was in the fields of economics and politics. As they were filtered through the philosophy of his friend Locke, and his successor Hume, Newton's ideas helped create the general scepticism of authority and belief in *laissez-faire*, which were to lower the prestige of religion and respect for a divinely constituted order of society. And so, rather unwittingly, Newton became one of the architects of the Enlightenment, whose focus was France.

Alien invasion: *Gulliver's Travels* and *Micromégas*

Newton's ideas were first introduced to the French through the prolific writings of François-Marie Arouet, better known by the pen name Voltaire. He was responsible for more than twenty thousand letters and two thousand books and pamphlets, with works in almost every literary fashion, including plays, poetry, novels, and essays. Voltaire was a free-thinker, and an outspoken champion of social reform. As a satirical polemicist, he also used distancing techniques in his literary works, in order to comment as an outsider on aspects of Western culture, a method that had been made popular by satirists such as Jonathan Swift.

In 1752, Voltaire published his *Micromégas*, a short story in which Earth is visited by two extraterrestrials, one from Saturn, the other from one of the planets of Sirius. The two alien visitors are not hostile to Earthlings. Indeed, they are somewhat enthused with benevolent curiosity, as they are engaged upon a 'short philosophical tour'[81] of the local cosmos, Earth included. But Voltaire had a hard act to follow.

The notion of alien invasion had only just been invented. In 1726, Jonathan Swift had published *Gulliver's Travels*, later named by English novelist George Orwell as one of the six most indispensable books in world literature.[82] Swift's novel was a skit on science and human nature, a parody of the contemporary traveller's tales. Like *Micromégas*, *Gulliver's Travels* was a quintessential illustration of the eighteenth-century planetary novel, which was often satirical in form.

Swift's 'flying island' of Laputa was fiction's first spaceship. But it was also literature's first parody of Newtonianism. The Laputian culture is a space colony of technologically superior 'people', a higher intelligence. The 'race of mortals' on Laputa is the most alien in all of Gulliver's travels. Heads cocked, and one eye looking inward as the other scans the zenith, the intellect of these madcap Newtons is constantly occupied. Draped in garments adorned with suns, moons and stars, they absent-mindedly speculate on mathematics and astronomy. These semi-crazed researchers, the first sighting of the mad scientist in fiction, have to be roused from their reverie by means of a rattle. Understandably their wives have run off.

In developing the idea of alien contact, Swift and Voltaire swam against the stream. The off-Earth tales of Godwin and de Bergerac had lowly Earthlings meeting up with powerful extraterrestrial civilisations. Following such pioneers, the majority of eighteenth-century planetary novels were voyages of human conquest in space. But Swift and Voltaire have the aliens come to Earth. Swift's lofty and tyrannical Laputians are akin to the brutal aliens of H.G. Wells' *The War of the Worlds* (1898), and countless other pulp fictions of the future. Voltaire's visitors are the benevolent guardians of Steven Spielberg's *Close Encounters of the Third Kind* (1977).

Swift had conjured up the diabolical alien. The 'flying island' of Laputa dominates the country above which it soars. Like some of the more infamous spaceships of twentieth-century fiction, any protest is punished. By manoeuvring Laputa, the land beneath can be deprived of sunshine and rain. A further reckoning may be exacted. Any discontents can have their towns subject to missile attack, or razed entirely, by having Laputa itself come crashing down to Earth. The ominous presence of this superior civilisation above the Earth is an early example of the fear of an alien menace. But it is also a satire on the inhumanity of science.

The Laputians are extreme Newtons. Every aspect of life on Laputa is run by science and mathematics. The beauty of an object is judged only by its geometry. Meat is dished up in geometric shapes. Clothes are tailored by compass and quadrant. Swift is clearly mocking the science of his day. Its redundancy and irrelevance is ridiculed. Its failure to create a practical payback lampooned. But the scientists of Laputa are nonetheless superior to their earthly counterparts. Gulliver represents a world, Newton's world, remedial in science, particularly astronomy and mathematics. The Laputians show little curiosity in the affairs of Earth, save a single terrestrial member of their court who is viewed as 'the most ignorant and stupid person among them'.

Voltaire was well aware of Swift's narrative of flying islands and mad scientists. *Gulliver's Travels* was an enormously influential book, and sublime in the effortless way in which it conjured up other, alien cultures. With Voltaire we have the other typical modern treatment of the cosmic invasion theme. Indeed, the treatment may well have its very provenance in *Micromégas*. Although Voltaire's is a benevolent type of encounter, it delves far further into man's identity than Swift's tale.

The mission of *Micromégas* is natural philosophy. As one of the leading Newtonian writers of his century, Voltaire was concerned with presenting the *philosophia naturalis*, whose values Newton had founded for the age. Voltaire had studied the cosmology of Newton and his disciples for some years, and published a popular form of the work in his *Éléments de la Philosophie de Newton* (1738). Voltaire was not so concerned with the physical aspects of Newton's system of the world. Rather, as a literary writer, he was fascinated by the implications of Newton's system for the new cosmos.

Micromégas is a young man from a planet in the vicinity of Sirius. He stands eight leagues high,[83] his size being in proportion to his planet of origin. Indeed, and by the same laws of proportionality argued by Huygens, Micromégas speaks of other aliens who are taller still, 'a single foot of whom is larger than this whole globe on which I have alighted'.[84] Similarly, the Saturnian who accompanies Micromégas on his philosophical tour is also in proportion with the diameter of his planet, standing at a mere six thousand feet high.[85] Together, the aliens are creative in their use of gravity. They take a ride on sunbeams and comets, presently reaching Earth at the northern coast of the Baltic Sea. They then embark upon a ramble around the Earth, a walk that takes a mere thirty-six hours.

At first, the extraterrestrials are sure that Earth is uninhabited. They are not able to see any living creatures. But fate is able to lend a helping hand. A diamond necklace worn by the miles-high Sirian snaps, scattering the diamonds to the ground. On collection, Micromégas notices that the crystals act as magnifiers, and with their help he spies first a whale and then a ship. Is the ship a beast? Once the aliens realise the ship has passengers, they catalogue the teeming inhabitants as 'mites' and are stunned that such tiny creatures should exist.

The prospect that these little 'atoms' of creatures could enjoy language, reason, and even possess a soul is regarded as risible. Until, that is, Micromégas is able to deftly raise the ship onto his thumbnail and engage in a somewhat faltering conversation with the little atoms, via an ear trumpet. But even as they manage to interpret the confused buzz that passes for conversation among the 'mites', they conclude that the entire affair is a bizarre 'jest of Nature'.[86]

Man was midway between telescope and microscope. As both instruments were employed as means of discovery in the pluralist cause, French writers, such as Fontenelle and Cyrano, used the idea of remote and hidden worlds, large and small, to evoke a cosmos replete with other possibilities. In *Micromégas*, Voltaire too stirs up the remotely small, and creates a comic variation on the merely relative view of man – placed as he was assumed to be, between the infinitely great and the infinitely small. The perspective was common in an age when the telescope and microscope vied with one another to reveal unimaginable worlds.

Voltaire also questions the idea of man as the peak of Creation. He stresses the very contemporary nature of the debate by using a scientific language and context of such large and small discoveries:

> What wonderful skill must have been inherent in our Sirian philosopher that enabled him to perceive those atoms of which we have been speaking. When Leuwenhoek and Hartsoeker observed the first rudiments of which we are formed, they did not make such an astonishing discovery.[87]

And so, Voltaire's intention becomes clear. Not only does this discovery of Micromégas undermine man's assumed midway position between the two voids, great and small; it also demotes man as the crowning of Creation.

Man, it seems, is not the measure of all things. How can he be? After all, man, this mere 'mite', is 'so near akin to annihilation'[88] that he is scarcely visible on the thumbnail of Micromégas. Even with the help of a makeshift microscope, in the form of a magnifying crystal, man is civilly pitied for his size, so small he is even asked if he has a soul. And yet this diminutive creature has the gall to pontificate about heaven and Earth. He boasts conclusive proof on the nature of the cosmos, even though there seem to be as many theories as thinkers.

And so Voltaire mocks the geocentric conceit. But Micromégas and the Saturnian are not narrow-minded. They admit that creatures smaller than man may be blessed with reason greater than that of the biggest giants of the visitors' grand cosmic tour. And in this vast scheme of things, any argument based on gradation of scale is now deemed ridiculous. Nonetheless, the visitors *are* scathing on man's arrogant anthropocentrism. Here is a creature, they declare, weakened by his worm's-eye perspective, who utters the most contradictory views about the nature of life in the universe. To hammer home the point, Voltaire has the ship peopled by delegates of all schools of philosophy: a ship of fools. *Micromégas* ends in a fit of 'inextinguishable laughter'.[89] A theologian from the Sorbonne tells them, 'their persons, their [worlds], their suns and their stars were created solely for the use of man',[90] as the aliens break into gales of laughter. Hubris is Voltaire's target here, especially the typically theological flavour illustrated by the Sorbonne priest.

And so, a full century after Cyrano, Voltaire takes pluralism a step further. In the name of the pluralist cause, Voltaire, like Cyrano, pokes fun at anthropocentric pride. Once Micromégas' laughter has died down, the gist of the tale becomes clear. Voltaire takes aim at the Church's anthropocentric teleology, as Micromégas promises the arguing terrestrial theologians his own book of wisdom, from which they will uncover the 'purpose of existence'.[91] His promise is kept, and the book is presented. It contains nothing but empty pages.

The climax of *Micromégas* suggests life itself is another 'jest of Nature'. Voltaire proposes, perhaps not without a wry grin, that the ultimate purpose of existence is unknown, and maybe even unknowable. In much of Voltaire's writing the same story is true. He seems to justify pluralism by analogy, rather than support it outright.[92] And especially not using the argument by design, which was so popular at the time. But if there is no divine and guiding hand, one might ask what Voltaire has to say about the meaning of a plurality of worlds in the cosmos.

Voltaire's is a plurality of humankinds. Paradoxically, by bringing the alien Micromégas down to Earth, Voltaire projects terrestrial troubles into the cosmos at large. Alien humans are neither better nor worse off than us. Micromégas himself had to leave his home planet because his scientific book had been banned by a censor who had not read it. Wherever you roam in the cosmos, life is much the same. For humans on other worlds, life is just as brief, love just as inconstant (though an erotic idyll on Saturn may last for a century), and other planetary grasses always greener.

But Voltaire's cosmos spells trouble for God. The problem is this. If the overriding human condition on all worlds is the same mundane struggle, the same unhappiness, the same irrational behaviour from beings who think themselves rational, then surely the justice of the Creator is in doubt. This problem with the philosophy of plurality had first been figured by St Thomas Aquinas. These diverse worlds cannot all be perfect. But neither can they all be identical, as Huygens had suggested in his own brand of pluralism.

Micromégas is a mystery. It leaves so many questions unanswered. Voltaire fails to comply with his pledge to unveil, at the end of his tale, the purpose of life in the universe. There seems to be no purpose on the part of the 'author of nature'.[93] Like Cyrano, Voltaire wonders why so many planets in the deep and mysterious cosmos should be so similar to Earth. What is the point of a plurality of worlds if they are no better than here. It is a tyranny of the tedious. The same warfare, the same censorship, with no word of a world being any better than another.

Finally, there is Micromégas' book on existence. Perhaps the book of empty pages symbolises the hope of revelation.[94] Or perhaps it represents the tyranny of reason. The universe at large may well be absurd; the void being totally devoid of any meaning. English poet John Milton had written about the 'vast unbounded Deep' unveiled by the telescope, reflecting the end of the medieval walled cosmos:

> Before [his] eyes in sudden view appear
> The secrets of the hoary Deep – a dark
> Illimitable ocean, without bound,
> Without dimension ...[95]

And in 1799, English poet William Wordsworth had written of Newton:

> Newton with his prism and silent face,
> The marble index of a mind for ever
> Voyaging through strange seas of Thought, alone.[96]

A fearful vision had been confronted. The argument by design was no longer fully trusted, and God's hand may not be present in the creation of a plurality of worlds. But an almost apocalyptic vision of the cosmos lay yawning and pregnant in its place. Such was the philosophically deep and challenging space literature of the eighteenth century. We have come far from the relatively naïve notions of the utopian space novels of Godwin and Kepler, which are nonetheless the precursors of Swift and Voltaire.

So the space stories had developed into affirmations of the new science. The nature of the philosophical conclusions of a tale such as Voltaire's *Micromégas* is of Newton's proof and completion of the Copernican revolution. As the eighteenth century bled into the nineteenth, the plurality of worlds debate intensified, in philosophy and in faith. The major figure to dominate the coming age was English polymath William Whewell, whose unitary Earth provided a robust challenge to the astronomers. But for science, one thing was for sure. The Copernican–Newtonian system pointed undeniably towards a pluralist future.

Notes

1 de Bergerac, C. (1965) *The Comical History of the Moon* in *The Other World or the States and Empires of the Moon*, trans. G. Strachan, Oxford
2 Randall, J. H. (1962) *The Career of Philosophy, Vol.* **1***: From the Middle Ages to the Enlightenment*, New York
3 Westfall, R. (1980) *Never At Rest: A Biography of Isaac Newton*, Cambridge
4 Quoted in Dick, S. J. (1982) *Plurality of Worlds: The Origins of the ET Life Debate, from Democritus to Kant*, Cambridge
5 ibid.
6 ibid.
7 ibid.
8 ibid.
9 Clarke, A. C. (2000) *Greetings, Carbon-Based Bipeds*, London
10 de Bergerac, C. (1971) *The Comical History of the Moon* in *The Other World or the States and Empires of the Moon*, London
11 ibid.
12 The shrine of Dodona was regarded as the oldest Hellenic oracle, possibly dating to the second millennium BC, according to Herodotus.
13 de Bergerac, C. (1971) *The Other World or the States and Empires of the Sun*, London
14 ibid.
15 ibid.
16 ibid.
17 Bernard le Bovier de Fontenelle (1686) *Entretiens sur la pluralité des mondes (Conversations on the Plurality of Worlds)*, ed. N. Rattner Gelbart, trans. H. A. Hargreaves (1990) Berkeley
18 Crowe, M. J. (2008) *The Extraterrestrial Life Debate, Antiquity to 1915, A Source Book*, Notre Dame
19 Almond, P. C. (2006) Adam, pre-Adamites, and extra-terrestrial beings in early modern Europe, *Journal of Religious History*, **30** (2): 163–74
20 Fontenelle was clearly a man of ambition. It did him no harm; he lived until he was one month short of his hundredth birthday.
21 Bernard le Bovier de Fontenelle (1686) *Entretiens sur la pluralité des mondes (Conversations on the Plurality of Worlds)*, ed. N. Rattner Gelbart, trans. H. A. Hargreaves (1990) Berkeley
22 ibid.
23 ibid.
24 ibid.
25 ibid.
26 ibid.
27 ibid.
28 ibid.
29 ibid.

30 ibid.
31 ibid.
32 ibid.
33 Galilei, G. (1610) *Sidereus Nuncius*, Venice
34 Bernard le Bovier de Fontenelle (1686) *Entretiens sur la pluralité des mondes* (*Conversations on the Plurality of Worlds*), ed. N. Rattner Gelbart, trans. H. A. Hargreaves (1990) Berkeley
35 ibid.
36 ibid.
37 The Accademia dei Lincei, literally the 'Academy of the Lynx-Eyed', and also known as the Lincean Academy, is an Italian science academy, founded in 1603 by Federico Cesi. It was the first academy of sciences to persist in Italy and a locus for the incipient Scientific Revolution.
38 Galileo had called it the 'occhiolino' or 'little eye'.
39 Bernard le Bovier de Fontenelle (1686) *Entretiens sur la pluralité des mondes* (*Conversations on the Plurality of Worlds*), N. Rattner Gelbart, trans. H. A. Hargreaves (1990) Berkeley
40 de Bergerac, C. (1965) *The Comical History of the Moon* in *The Other World or the States and Empires of the Moon*, trans. G. Strachan, Oxford
41 Quoted in Crowe, M. J. (2008) *The Extraterrestrial Life Debate, Antiquity to 1915, A Source Book*, Notre Dame, p. 86
42 ibid.
43 Huygens, C. (1698) *The Celestial Worlds Discover'd: or, Conjectures Concerning the Inhabitants, Plants and Productions of the Worlds in the Planets*, London
44 ibid.
45 ibid.
46 ibid.
47 ibid.
48 ibid.
49 ibid.
50 ibid.
51 ibid.
52 ibid.
53 ibid.
54 ibid.
55 Kuhn, T. (1957) *The Copernican Revolution: Planetary Astronomy in the Development of Western Thought*, Harvard
56 *Oeuvres des Descartes*, ed. C. Adams and P. Tannery, vol. V (Paris, 1903), pp. 54–5
57 *Philosophiæ Naturalis Principia Mathematica* (*Mathematical Principles of Natural Philosophy*)
58 Isaac Newton, *Mathematical Principles of Natural Philosophy*, passage trans. W. H. Donahue, in M. J. Crowe (2007) *Mechanics from Aristotle to Einstein*, Santa Fe, pp. 190–93
59 Newton, I. (1952) *Opticks: Fourth Edition*, New York, pp. 403–4
60 Whiston, W. (1717) *Astronomical Principles of Religion*, London, pp. 91–7
61 Pope, A. *Essay on Man*, Epistle I, lines 23–8
62 As quoted in Hebb, R. C. (1882) *Bentley*, New York
63 *Four Letters from Sir Isaac Newton to Doctor Bentley* (London, 1756), reprinted in *Isaac Newton's Papers and Letters on Natural Philosophy*, ed. I. Bernard Cohen (1958) Cambridge, Mass

64 ibid.
65 Bentley, R. (1958) *A Confutation of Atheism* in *Isaac Newton's Papers and Letters on Natural Philosophy*, ed. I. Bernard Cohen, Cambridge, Mass
66 *Newton's Papers and Letters*, p. 356
67 *Newton's Papers and Letters*, p. 358
68 *Newton's Papers and Letters*, p. 359
69 *Newton's Papers and Letters*, p. 368
70 Derham, W. (1715) *Astro-Theology*, 2nd edn., London, p. xli
71 ibid., p. xliv
72 Brewster, D. (1855) *Memoirs of Life, Writings, and Discoveries of Sir Isaac Newton*, vol. II, Edinburgh, p. 354
73 Quoted in Crowe, M. J. (1986) *The Extraterrestrial Life Debate, 1750–1900*, New York, p. 25
74 Quoted in Crowe, M. J. (2008) *The Extraterrestrial Life Debate, Antiquity to 1915, A Source Book*. Notre Dame, p. 114
75 ibid.
76 Crowe, M. J. (1986) *The Extraterrestrial Life Debate, 1750–1900*, New York, p. 25
77 ibid.
78 ibid.
79 ibid.
80 Bernal, J. D. (1965) *Science in History*, London, p. 488
81 Wade, I. O. (1950) *Voltaire's Micromégas: A Study in the Fusion of Science, Myth, and Art*, Princeton, p. 127
82 Orwell, G. (1968) *Politics vs. Literature: An Examination of Gulliver's Travels* in *The Collected Essays, Journalism and Letters of George Orwell*, London
83 or 24 000 geometrical paces of five feet each, or 120 000 statute feet
84 Wade, I. O. (1950) *Voltaire's Micromégas: A Study in the Fusion of Science, Myth, and Art*, Princeton, p. 139
85 ibid., p. 122
86 ibid., p. 137
87 ibid., p. 135
88 ibid., p. 138
89 ibid., p. 146
90 ibid., p. 145
91 ibid., p. 146
92 Guthke, K. S. (1990) *The Last Frontier: Imagining Other Worlds from the Copernican Revolution to Modern Science Fiction*, New York, p. 304
93 Wade, I. O. (1950) *Voltaire's Micromégas: A Study in the Fusion of Science, Myth, and Art*, Princeton, p. 125
94 Guthke, K. S. (1990) *The Last Frontier: Imagining Other Worlds from the Copernican Revolution to Modern Science Fiction*, New York, p. 306
95 Milton, J. (1667) *Paradise Lost*, book ii, l. 890
96 Wordsworth, W. (1799) *The Prelude*

4

Extraterrestrials in the early machine age

It seemed to the ancients that there was only one Earth inhabited, and even of that men held the antipodes in dread: the remainder of the world was, according to them, a few shining globes and a few crystalline spheres. Today, whatever bounds are given or not given to the Universe, it must be acknowledged that there is an infinite number of globes, as great as and greater than ours, which have as much right as it to hold rational inhabitants, though it follows not at all that they are human.

It is only one planet, that is to say one of the six principal satellites of our Sun; and as all fixed stars are suns also, we see how small a thing our Earth is in relation to visible things, since it is only an appendix of one amongst them. It may be that all suns are peopled only by blessed creatures, and nothing constrains us to think that many are damned, for few instances or few samples suffice to show the advantage which good extracts from evil.

Moreover, since there is no reason for the belief that there are stars everywhere, is it not possible that there may be great space beyond the region of the stars? Whether it be the Empyrean Heaven, or not, this immense space encircling all this region may in any case be filled with happiness and glory. It can be imagined as like the Ocean, whither flow the rivers of all blessed creatures, when they shall have reached their perfection in the system of the stars.

What will become of the consideration of our globe and its inhabitants? Will it not be something incomparably less than a physical point, since our Earth is as a point in comparison with the distance of some fixed stars? Thus since the proportion of that part of the Universe which we know is almost lost in nothingness compared with that which is unknown, and which we yet have cause to assume, and since all the evils that may be raised in objection before us are in this near nothingness, haply it may be that all evils are almost nothingness in comparison with the good things which are in the Universe.

Gottfried Wilhelm Leibniz, *Théodicée*, trans. E.M. Huggard[1]

The machine age, man-bats, and the Great Moon 'Hoax'

By the dawn of the nineteenth century the early vision of material progress through science had been achieved. Science had secured its dominion over nature. Newtonianism had established its authority in the clanging new workshop of the world that was Victorian Britain. 'Were we required', wrote Thomas Carlyle in 1829, 'to characterise this age of ours by any single epithet, we should call it the Mechanical Age'.[2]

Newton's system of the world was made flesh.

The *philosophical* engine, the early steam engine, drove trains along their metal tracks; the first gleaming steamships crossed the oceans; transport magnates were building great bridges and roads; telegraphs ticked intelligence from station to station; cotton works glowed by gas; and a clamorous arc of iron foundries and coal mines powered this Industrial Revolution. Newton had conjured up a mechanical worldview, a clockwork cosmos. The process that had begun with the telescope and the microscope marched on relentlessly.

As the machinery began to mesh, science encroached upon all aspects of life, all corners of the globe, and beyond. The machines of science were devised not merely to explore, but to exploit. And one of the most rapidly expanding areas of industry was newspaper publication. By the early nineteenth century there were over fifty newspapers in London alone, and another hundred titles in Britain. And as tax duties were reduced from the 1830s onwards, there was a massive explosion in circulation, as the newspapers fed public hunger for communication. In post-revolutionary United States, the story was the same. Publications spread like wildfire. In 1800 there were up to two hundred titles, by 1810 over 350, and during the following two decades the increase was equally rapid.

In this climate emerged perhaps the most remarkably cosmic newspaper report ever published: the discovery of a civilisation on the Moon. The 'hoax' began on 25 August 1835 when the *New York Sun* carried the first of six instalments professed to be a lunar revelation by the eminent astronomer Sir John Herschel. So breathtaking was the detail of the initial report that over nineteen thousand copies of the 26 August issue of the *New York Sun* were sold. It was the largest circulated newspaper on the planet.

The publishers smelled success, and resounding sales. The 29 August issue of the *Sun* declared that the entire report was to be sold as a pamphlet. Within days sixty thousand copies of the pamphlet had been sold. And the booklet has made a number of re-appearances ever since. Illustrations, such as the lithograph in Figure 10, of the Lunarians were also sold. And various translations of this extraordinary publication popped up around the globe. Even by the close of 1836 there had been Italian translations in Florence, Naples, Ravenna, and

Figure 10 Great Moon 'Hoax' lithograph depicting man-bats in a 'ruby amphitheatre', from *New York Sun*, 28 August 1835 (article 4, of 6).

Livorno; French translations in Paris, Strasbourg, Lyon, Bordeaux, and Lausanne; a German translation in Hamburg; and Spanish translations in Mexico and Cuba.

This much of the 'hoax' was true. In 1835, English astronomer John Herschel *was* at the Cape of Good Hope, as the newspaper reports suggested. But the publication from which the report was allegedly culled, namely the *Edinburgh Journal of Science*, was no longer in print. Nonetheless, the article preview gave a hint of what was to come:

> by means of a telescope, of vast dimensions and an entirely new principle, the younger Herschel ... has already made the most extraordinary discoveries in every planet of our solar system; has obtained a distinct view of objects in the Moon ... has affirmatively settled the question whether this satellite be inhabited, and by what order of beings; has firmly established a new theory of cometary phenomena; and has solved or corrected nearly every leading problem of mathematical astronomy.[3]

The *New York Sun* kept its readers waiting. For the most part, the first instalment was a mere description of Herschel's telescope, which boasted a twenty-four-foot mirror capable of forty-two thousand times magnification. But with the second issue on 26 August, the 'observed' lunar world was unveiled. First, a tantalising glimpse of the Moon's geology and botany, and then Herschel's scribe, Dr Andrew Grant, declares, 'our magnifiers blest our panting hopes with specimens of conscious existence'.[4] They spy a being like a bison, which is seen to have

> one widely distinctive feature, which we afterwards found common to nearly every lunar quadruped we have discovered; namely, a remarkable fleshy appendage over the eyes, crossing the whole breadth of the forehead and united to the ears. We could most distinctly perceive this hairy veil, which was shaped like the upper front outline of the cap known to the ladies as Mary Queen of Scot's cap, lifted and lowered by means of the ears. It immediately occurred to the acute mind of Dr Herschel that this was a providential contrivance to protect the eyes of the animal from the great extremes of light and darkness to which all the inhabitants of our side of the Moon are periodically subjected.[5]

Their delight knows no bounds as they spot a bearded hircine beast, blessed with a prominent horn. No sooner have they spied a flock of stunning birds and a shoal of lunar fish than the instalment ends with the setting of the Moon. The next issue of 27 August kept the readers waiting, with reports of lunar volcanoes and seas, but the jackpot came the following day when Herschel and his co-observers reported seeing creatures who

> averaged four feet in height, were covered, except on the face, with short and glossy copper-coloured hair, and had wings composed of a thin membrane, without hair, lying snugly upon their backs, from the top of the shoulders to the calves of the legs. The face, which was of a yellowish flesh-colour, was a slight improvement upon that of the large orang-outang, being more open and intelligent in its expression, and having a much greater expansion of forehead ... In general symmetry of body and limbs they were infinitely superior to the orang-outang; so much so, that, but for their long wings, Lieut. Drummond said they would look as well on a parade ground as some of the old cockney militia![6]

The question of Lunarian intelligence was swiftly solved when the creatures are observed

> evidently engaged in conversation; their gesticulation, more particularly the varied action of their hands and arms, appeared impassioned and emphatic. We hence inferred that they were rational beings, and, although not perhaps of so high an order as others which

> we discovered the next month on the shores of the Bay of Rainbows, that they were capable of producing works of art and contrivance.[7]

The creatures are then christened 'man-bat', and gifted with the accolade of intelligence: 'We scientifically denominated them the Vespertilio-homo, or man-bat; and they are doubtless innocent and happy creatures, notwithstanding some of their amusements would but ill comport with our terrestrial notions of decorum.'[8] When the 29 August issue was published, *Sun* readers were treated to reports of lunar oceans and a 'magnificent... temple – a fane of devotion, or of science, which when consecrated to the Creator, is devotion of the loftiest order'.[9] The series of reports was brought to an end with the last instalment, on 31 August, climaxing in a high point – a description of yet higher creatures and a telling of the 'universal state of amity among all classes of lunar creatures'.[10] A brief appearance is made by Saturn, but Herschel's calculations are rather patronisingly left out of the report, 'as being too mathematical for popular comprehension'.[11] As the series is brought to an end, the highest species of man-bat is spied, 'In stature, they did not excel those last described, but they were of infinitely greater personal beauty, and appeared, in our eyes, scarcely less lovely than the general representation of angels by the more imaginative school of painters.'[12]

The United States was abuzz with the 'discovery'. On 1 September, the *New York Sun* reported on the reactions of other publications in the state to the lunar revelations. The *Mercantile Advertiser* declared, 'It appears to carry intrinsic evidence of being an authentic document'.[13] The *Daily Advertiser* triumphantly suggested, 'No article, we believe, has appeared for years, that will command so general a perusal and publication. Sir John has added a stock of knowledge to the present age that will immortalize his name, and place it high on the page of science,'[14] The *New York Times*, according to the *Sun*'s account, decreed the discoveries 'probable and plausible',[15] while the *New Yorker* decided they ushered in 'a new era in astronomy and science generally'.[16]

Predictably, the church was not entirely amused. One contemporary account speaks of 'Some of the grave religious journals made the great discovery a subject of pointed homilies'.[17] Another report tells of an American clergyman who forewarned his congregation that he would start up a fighting fund for Bibles to donate to the lunar, and no doubt heathen, bat-people. Even the question of slavery cropped up. The Slavery Abolition Act had been passed through the UK Parliament in 1833, and had abolished slavery throughout most of the British Empire. During the Great Moon 'Hoax', it was claimed, 'the philanthropists of England had frequent and crowded meetings at Exeter Hall, and appointed committees to inquire ... in regard to the condition of the people of the Moon, for purposes of relieving their wants, ... and, above all, abolishing slavery if it should be found to exist among the lunar inhabitants.'[18]

There were eyewitness reports of the 'hoax'. It seems some academics were not sufficiently sceptical of the *Sun* coverage, 'Yale was alive with staunch supporters. The literati – students and professors, doctors in divinity and law – and all the rest of the reading community, looked daily for the arrival of the New York mail with unexampled avidity and implicit faith.'[19] Eminent American author Edgar Allan Poe was later to report,

> Not one person in ten discredited it, and (strangest point of all!) the doubters were chiefly those who doubted without being able to say why – the ignorant, those uninformed in astronomy, people who *would not* believe because the thing was so novel, so entirely 'out of the usual way'. A grave professor of mathematics in a Virginia college told me seriously that he had *no doubt* of the truth of the whole affair.[20]

It was neither the first nor last cosmic controversy, of course. In Chapter 2 we spoke of the public controversy that followed the publication of Galileo's observations through the telescope. Here again was a controversy emanating from novel, if not Earth-shattering, observations. And the same drama would be played out again with the Martian canals controversy in the late nineteenth century, and the UFO farce of the mid twentieth.

At last, though, the bubble burst on the Great Moon 'Hoax'. A journalist from the *Journal of Commerce* was dispatched to the *New York Sun*. His mission: to secure a copy of the full Moon report so that the *JoC* may consider re-publication. The reporter met with a *Sun* staff writer by the name of Locke, who told him, 'Don't print it right away. I wrote it myself.'[21] The *JoC* subsequently accused the *Sun* of publishing a hoax, and the *New York Herald* named Locke as the architect.

Pluralism in the machine age

And so Richard Adams Locke finally enters the frame. Locke was an American descendant of the Newtonian philosopher, John Locke. Some accounts have Richard Adams Locke studying at the University of Cambridge, though Cambridge has no evidence he ever did so. He did, however, work as a writer and editor in England before setting up in New York, where eventually he joined the *Sun* in 1835. According to several biographies, Locke seems to have been interested in science, which is resoundingly confirmed by the impressive detail about pluralism in his *Sun* articles. Edgar Allan Poe had Locke down as a skilful writer. Just prior to 1835, Poe had written his own lunar fiction, so he was well placed to comment on Locke's talent in writing with 'true imagination'.[22]

Professor Michael J. Crowe has made a compelling case that the Great Moon 'Hoax' was in fact merely satire.[23] And he presents 'solid historical evidence'[24] that Locke uses his science communication skills at the 'delicate task of writing satire'.[25] Locke's target was the pluralism of the early machine age, and Thomas Dick in particular. Dick had been a Scottish churchman, and also a science teacher and writer. He was well known for his works on astronomy, as well as his attempts to defuse tension between the two fields.

Not everyone was pleased by Dick. In his 1852 account of the 'hoax', William Griggs was critical of Dick's work:

> it would be difficult to name a writer who, with sincere piety, much information, and the best of intentions, has done greater injury, at once, to the cause of rational religion and inductive science, by the fanatical, fanciful, and illegitimate manner in which he has attempted to force each into the service of the other, instead of leaving both to the natural freedom and harmony of their respective spheres.[26]

The Newtonian truce is writ large in this above quote. In the last chapter we spoke of the unwritten compromise that science had made, to reject the more radical aspects of materialist Atomism, and not to trespass into the field of religion. So it was with Griggs' sensibilities. But he had no doubt that Dick was the subject of Locke's infamous spoof:

> we have the assurance of the author, in a letter published some years since, in the New World, that it was written expressly to satirise the unwarranted and extravagant anticipations upon this subject, that had been first excited by a prurient coterie of German astronomers, and thence aggravated almost to the point of lunacy itself, both in this country and in England, by the religio-scientific rhapsodies of Dr Dick. At that time the astronomical works of this author enjoyed a degree of popularity, in both countries, almost unexampled in the history of scientific literature. The scale of the editions republished in this country was unbounded until nearly the whole of his successive volumes found a place in every private and public library in the land.[27]

Indeed, Griggs pressed the point even further. He confirmed that the articles had been penned as a result of Locke's reading material during the summer of 1835. He had been reading an 1826 issue of the *Edinburgh New Philosophical Journal*. And in this issue Dick had reported the idea of constructing vast geometrical signals, by which Earthlings could communicate with 'selenites', as the assumed inhabitants of our Moon were known. The strikingly unusual nature of this early idea of interspecies communication bears a closer look.

The architect of the selenite signals was Carl Friedrich Gauss. Gauss was something of a colossus in nineteenth-century science. Sometimes known as the 'Prince of Mathematicians', and even the 'greatest mathematician since antiquity', he was also one-time director of the Göttingen Observatory in Germany. Gauss appears to have been a pluralist, and a believer in lunar life, a belief that he was allegedly keen to pursue in a scientific culture that was increasingly sceptical about evidence of life on the lunar surface.

Gauss' plan was this: send signals to the Moon or Mars, so that extraterrestrials would know we were smart. His preferred methods of sending such signals were ambitious and dramatic. Method one: in the Siberian forests, set up a vast 'windmill' diagram as used in Euclid's illustration of Pythagoras' theorem. Naturally, the method assumed, the selenites would respond in kind, as mathematics on the Moon was just the same as that on Earth. The lunar men would realise Earth was inhabited and draw fitting conclusions.

Method two was no less dramatic: the construction of a huge canal, cut into the Saharan desert. Once the canal had been cut the climax of the master plan would be realised: onto the canal waters paraffin[28] would be poured, and the whole thing set ablaze. Using this method, a different signal could be sent every night. Once more the selenites would be stirred by such a striking display of terrestrial intelligence. This second method seems to have been the brainchild of a colleague of Gauss, one Johann Joseph von Littrow, the director of the Vienna Observatory.

Alas, as with most things associated with Richard Adams Locke, the selenite signal story appears to be apocryphal. Not only is there many a different telling of the tale, but also no narrative actually contains reference to the writings in which Gauss and Littrow are originally meant to have set down their respective plans of lunar action. And yet the account acts as a perfect introduction to the wild and wonderful world of Thomas Dick. Indeed, the sheer potential farce of the ideas led Locke to regard the science as 'fair subjects of sedate and elaborate satire',[29] as Griggs was to quote.

But the signals story was just a beginning. Locke then opened the works of Thomas Dick himself, and found it hard to believe what he was reading. One thing was for sure, however, the Dick back catalogue was ripe for lampooning. Griggs was not alone in considering Locke's articles satirical. At the French Académie des Sciences, astronomer François Arago had read them aloud to other academy members among 'repeated interruptions from uncontrollable and uproarious laughter'.[30] John Herschel had mixed feelings. At first he laughed along with those in the know. But later, after receiving questions from believing English, German, French, and Italian enquirers, the farce was not so funny.

There appears to have been some academic debate on whether Locke's 'hoax' was indeed a satire. But this was finally laid to rest in 2004 when James Secord, a prominent science historian at Cambridge, located the original letter, dated 6 May 1840, that Locke sent to Griggs explicitly saying the articles were satirical. Locke wrote that he 'had become convinced that the *imaginative school of philosophy* . . . was emasculating the minds of our studious youth', and weakening them for normal science. He adds,

> One of the most conspicuous of the jingling heads of this school, is the famous Dr Dick of Dundee, who pastes together so many books about the Moon and stars, and devoutly helps out the music of the spheres with the nasal twang of the conventicle. It was this ciphering sage's '*Christian Philosopher*' that suggested the Moon-story.[31]

Ironically, Locke's satire failed. And yet that very fact is telling. Locke had miscalculated the mood of a gullible generation raised on the works of Thomas Dick, Thomas Chalmers, and their pluralist disciples. As one writer was to put it in 1852, 'The soil had been thoroughly ploughed, harrowed and manured in the mental fields of our wiser people, and the seed of farmer Locke bore fruit a hundred fold.'[32]

The Great Moon 'Hoax' is an entertaining and fitting introduction to pluralism in the machine age. Locke's satire illuminates the problems for pluralists in the early nineteenth century, as they were caught on the horns of a dilemma: religion on one side, and lack of scientific evidence on the other. Locke's spoof uncovered the naïveté of the pluralist readership. It exposed the overstated claims of its authors, and struck a much-needed note of caution in the debate. The 'hoax' was also one of the main turning points in the nineteenth-century pluralist debate. Another was the publication in 1853 of William Whewell's sceptical *Of the Plurality of Worlds: An Essay*. But before we weigh in with Whewell, let us consider the pluralist positions in the first half of the century.

The Caledonian canons

Among the fledgling century's most eminent pluralist champions were a couple of Scottish ministers. Both were at one with the age in that they endorsed the Enlightenment obsession with extraterrestrial life. One Caledonian canon we have already met in Thomas Dick. But before we delve further into Dick, we should consider the other churchman, who flirted with pluralism for just a few influential months: the Fife-born mathematician and divine Dr Thomas Chalmers.

It is typical of the scientific revival of this time that innovation should come not from the south of Britain, but from the north. The spirit of the new liberalism united free thought in science, manufactures, and *laissez-faire*. Indeed, there had been a Scottish renaissance in the eighteenth and nineteenth centuries. David Hume had provided a link with the philosophies of France; Adam Smith, with his *Wealth of Nations*, was the intellectual father of *laissez-faire* capitalism; Dr Joseph Black was the originator of the pneumatic revolution, one of his students being fellow Scotsman James Watt; and Dr James Hutton and Sir Charles Lyell were the founders of modern geology.

Thomas Chalmers had burst upon the scene at noon on Thursday 23 November 1815. The occasion was hardly auspicious. At the time, Chalmers was an obscure parson from rural Scotland, and he was about to embark on a series of seven weekday sermons, from the pulpit of Tron Church in Glasgow. And yet his sermons caused something of a sensation. To be sure, they acted as a springboard that launched the cometary Chalmers into orbit. The highlights of the career that followed included his recognition as the most important man in Scottish religion, and Thomas Carlyle's description of Chalmers as 'The Chief Scottish man of his time'.[33]

Witness this reaction to Chalmers from a contemporary eye-witness account of his series of sermons:

> One or two of these 'Discourses' ... were heard by the writer ..., then a boy. He had to wait nearly four hours before he could gain admission as one in a crowd in which he was nearly crushed to death. It was with no little effort that the great preacher could find his way to the pulpit. As soon as his fervid eloquence began to stream from it, the intense enthusiasm of the auditory became almost irrestrainable; and in that enthusiasm the writer, young as he was, fully participated. He has never since witnessed anything equal to the scene.[34]

Next came Chalmers in print. The sermons were collectively published early in 1817 as *A Series of Discourses on the Christian Revelation Viewed in Connection with the Modern Astronomy*. Again the reaction was stunning. The first ten weeks after publication saw the sale of six thousand copies. By the end of that year twenty thousand copies in nine editions had been purchased. Chalmers' biographer, William Hanna, said of the book in 1851, 'Never previously, nor ever since, has any volume of sermons met with such immediate and general acceptance ... It was, besides, the first volume of Sermons which fairly broke the lines which had separated too long the literary from the religious public.'[35]

Chalmers' book spread 'like wildfire through the country',[36] according to literary critic William Hazlitt. The enthusiasm continued unabated. By 1851, Chalmers' *Discourses* still represented the bulk of his sales. In Britain, America, and Europe, the sales of the book persisted late into that century. Hardly surprising, then, that noted American geologist, and third President of Amherst

College, Edward Hitchcock claimed in the 1850s, 'All the world is acquainted with Dr Chalmers' splendid *Astronomical Discourses*'.[37]

Often, of course, there is nothing more powerful in a story than the very telling of it. So it was with Chalmers. Even in the mean streets of Glasgow they feared the potent and evangelical enthusiasm of the young preacher. Chalmers had many imitators, but as one commentator put it, 'young men, and they were not few, who tried to ape his manner and apply his method became the bores of their generation'.[38] Bearing this in mind, it is reasonable to conclude that no amount of biographical detail will conjure up the same impact felt by Chalmers' flock in the nineteenth century.

And so to the book itself. Chalmers' *Discourses* is evangelical from the very start. He makes clear his intentions in the preface. Chalmers takes serious issue with (1) the 'assertion that Christianity is a religion which professes to be designed for the single benefit of our world',[39] and (2) the 'inference ... that God cannot be the author of this religion, for he would not lavish on so insignificant a field such peculiar and such distinguishing attentions as are ascribed to Him in the Old and New Testament.'[40] To the first assertion, Chalmers argues that such a contention is only that of the infidel, not of the Christian. And to the second assertion, he conjures up a cosmos presided over by a God of such grace and power that his goodness is almost without limit, his universe replete with life.

In the first sermon, the new Christian cosmos unfolds before the reader. The heavens detailed in the *Discourses* are those of the telescope, not of the psalmists. With instruments of discovery we have found planets that dwarf the Earth, other worlds with seasons and satellites, and stars in staggering number, many of which are home to their own planetary systems. Chalmers suggests that these planets 'must be the mansions of life and intelligence'.[41] and even where the telescope cannot spy, 'there are other worlds, which roll afar; the light of other suns shines upon them; and the sky which mantles them, is garnished with other stars.'[42]

After eulogising Newton in his second sermon, Chalmers sings the praises of 'scopes in the third. Here he suggests that the microscope, invented practically alongside the telescope, offers an answer, 'The one [the telescope] led me to see a system in every star. The other leads me to see a world in every atom.'[43] As the microscope reveals God's grace in even the smallest creatures, so will His beneficence extend to minds on other orbs. In the previous chapter, we reported Thomas Kuhn's account of the challenges facing Christianity in the new Copernican cosmos. These included questions as to whether Christ would have come as redeemer to other worlds, as he did to Earth. Chalmers has an emphatic answer:

> for any thing we can know by reason, the plan of redemption may have its influences and its bearings on those creatures of God who people

> other regions, and occupy other fields in the immensity of his dominions; that to argue, therefore, on this plan being instituted for the single benefit of . . . the species to which we belong, is a mere presumption of the Infidel himself.[44]

Aliens sing hosannas in the fifth sermon. After re-reading some Scripture as pluralist in the fourth sermon, Chalmers speaks of joy in heaven in the next. He paints a picture of the entire cosmos rejoicing at the repentance of one Earthling, he portrays them as ringing 'throughout all their mansions the hosannas of joy, over every one individual of [Earth's] repentant population'.[45] And he adds, 'For anything I know, every planet that rolls in the immensity around me, may be a land of righteousness.'[46] His conclusion is an arresting one: that the cosmos may be 'one secure and rejoicing family [wherein our] alienated world is the only strayed, or only captive member'.[47]

Satan pops up in the sixth sermon, and with him cosmic warfare. In a sermon entitled *Contest for an Ascendency over Man, amongst the Higher Orders of Intelligence*, Chalmers speculates whether 'our rebellious world be the only strong-hold which Satan is possessed of, or if it be but the single post of an extended warfare, that is now going on between the powers of light and darkness'.[48] Nonetheless, he explains, 'why on the salvation of our solitary species so much attention appears to have been concentrated'.[49] Finally, in the seventh and last sermon, he calls for a humble life of Christian commitment.

The *Discourses* of Chalmers was a curious wonder. The sermons were penned not by an astronomer, or even a theologian, but by a brilliant orator. Chalmers was an evangelical preacher, aiming to win the hearts, minds, and souls of the flock that came into his ken. Given the overwhelming reaction to his book, many readers must have become pluralists, and Christianity afforded a new dignity. Though full of fluency and rich imagery, Chalmers' sermons are flawed, of course. There is little scientific detail. One would be hard pressed to find where in the vast cosmos Chalmers believes extraterrestrial life to be.

The many worlds of Dr Dick

The pluralism of Thomas Chalmers was an emollient to the faint-at-heart faithful. Contemporary cosmology in the hands of scientists such as the German astronomer William Herschel had become a rather frightening affair. Herschel's vast telescopes evoked speculations on the life and death of the stars themselves. It seemed that nothing was eternal any more. In such a bewildering universe, the faithful took solace in the soothing and metaphoric imagery of Chalmers. And that is exactly the point.

Chalmers' work was a damage limitation exercise. Ever since Copernicus, astronomy had unveiled a universe decentralised, inhuman, and alien. The more recent and wilder speculations of astronomers William Herschel and Johann Elert Bode seemed weirder still. Herschel declared every planet to be inhabited, even the Sun, which had a cool, solid surface protected from its hot atmosphere by an opaque layer of cloud. According to Herschel, the Sun housed a race of creatures fittingly adapted to their alien environment. They possessed enormous heads, as his calculations showed that under such conditions normal heads would explode.

Bode's interests also extended to the Sun. He championed pluralism with great frequency and influence, and regarded the Sun's inhabitants, the Solarians, to benefit from a cool core, surrounded by a protective layer. Concerning these solar inhabitants he asked,

> Who would doubt their existence? The most wise author of the world assigns an insect lodging on a grain of sand and will certainly not permit . . . the great ball of the sun to be empty of creatures and still less of rational inhabitants who are ready gratefully to praise the author of their life. Its unfortunate inhabitants, say I, are illuminated by an unceasing light, the blinding brightness of which they view without injury and which, in accordance with the most wise design of the all-Good, communicates to them the necessary warmth by means of its thick atmosphere.[50]

To these notions of Herschel and Bode, the everyday Christian might be totally alienated. Chalmers responded in the sense that, as one commentator was to put it, 'he is seeking, not so much to prove theism, as to remove difficulties in the way of belief'.[51]

Thomas Dick was to go one step further. Whereas the doctrine of pluralism perturbed Chalmers for but a matter of months, the question of cosmic life was to loom large in the legend of Dr Dick from Dundee. The city was the seat of Dick's positive blitz on pluralism, in a series of books published between 1823 and 1840. The first of these, *Christian Philosopher*, was to establish his credentials in the field. It was followed by a bevy of further books, *Philosophy of Religion* (1826), *Philosophy of a Future State* (1828), *Celestial Scenery* (1837), and *Sidereal Heavens* (1840).

To get a taste of Dick, let us dip into his *Celestial Scenery*, in the last chapter of which he details five arguments for the plurality of worlds:

1. *There are bodies in the planetary system of such* MAGNITUDES *as to afford ample scope for the abode of myriads of inhabitants;*
2. *There is a* GENERAL SIMILARITY *among all the bodies of the Planetary System, which tends to prove that they are intended to subserve the same ultimate designs in the arrangement of the Creator;*

3. *IN the bodies which constitute the solar system, there are* SPECIAL ARRANGEMENTS *which indicate their* ADAPTION *to the enjoyments of sensitive and intelligent beings; and which prove that this was the ultimate design of their creation;*
4. *The scenery of the heavens, as viewed from the surfaces of the larger planets and their satellites, forms a presumptive proof that both the planets and their moons are inhabited by intellectual beings;*
5. *In the world we inhabit, every part of nature is destined to the support of animated beings.*[52]

Three years later, in his *Sidereal Heavens* (1840), Dick was to add a further three cases for pluralism:

6. *That the doctrine of a plurality of worlds is more worthy of the perfection of the Infinite Creator, and gives us a more glorious and magnificent idea of his character and operations than to suppose his benevolent regards confined to the globe on which we dwell;*
7. *Wherever one perfection of Deity is exerted, there also ALL his attributes are in operation, and must be displayed, in a greater or less degree, to certain orders of intelligences;*
8. *There is an absurdity involved in the contrary supposition – namely, that the distant regions of creation are devoid of inhabitants.*[53]

Clearly, Dick was evangelical about extraterrestrial life. In his *Christian Philosopher* (1823), he extols the virtues of divine wisdom that placed the Sun at just the right distance to 'refresh and cheer us, and to enliven our soil'.[54] But it is not just Earthlings that are so blessed. Inhabitants of the other planets also benefit from the same gladdening conditions, so much so that there are those that need no sleep. Even beyond the Solar System, Dick believes this case to be true, 'every star ... is the centre of a system of planetary worlds, where the agency of God may be endlessly varied, and perhaps, more strikingly displayed than even in the system to which we belong'.[55]

Dick even outdoes William Herschel. He suggests that God has created, within the Sun, 'a number of worlds ... and peopled them with intelligent beings'.[56] Dick discards Herschel's idea of volcanoes on the Moon, and suggests instead, 'It would be a far more pleasing idea, and perhaps as nearly corresponding to fact, to suppose that these phenomena are owing to some occasional splendid illuminations, produced by the lunar inhabitants, during their long nights.'[57] He prophesies that direct evidence of lunar inhabitants will soon be found, though casts doubt upon the alleged evidence to date.

Thomas Dick had at his disposal an adept command of both astronomy and Scripture. This put him at a distinct advantage in debate, though he was often guilty of running away with his speculations. For instance, though he claimed comets were without life, as their purpose was unknown, he was quite sure that the planetoids were populated. Supporting the idea that they had been formed from fragments of a larger body, Dick held that the explosion responsible 'would seem to indicate, that a moral revolution has taken place among the intelligent beings who had originally been placed in those regions'.[58] A rather curious conclusion to draw.

But then Dick was no stranger to bold and curious conclusions. In his *Philosophy of Religion* (1826), he proclaimed, 'The grand principles of morality . . . are not to be viewed as confined merely to the inhabitants of our globe, but extend to all intelligent beings . . . throughout the vast universe',[59] later adding, 'there is but one religion throughout the universe'.[60] He was equally bold with his calculations.

His *Philosophy of a Future State* (1828) included the boldest possible estimates about Dick's pluralist universe. He calculates that, since we can see eighty million stars and because each star has at least thirty bodies in orbit about it, 2.4 billion inhabited worlds must dwell in the *visible* universe alone. Once more mixing religion with science, Dick then goes on to claim that our immortal souls will spend much of eternity learning of the history and geography of these other worlds.

Not even God and the angels can escape his gaze. The angels are demoted from spiritual to material beings, as he speculates that celestial messengers will convey news to us on the other worlds. As for God, Dick holds the 'highly probable, if not certain'[61] idea that astronomers will one day locate, at the centre of the universe, the 'Throne of God'.[62] 'Here', he continues, 'deputations from all the principal provinces of creation, may occasionally assemble, and inhabitants of different worlds mingle with each other, and learn the grand outlines of those physical operations and moral transactions, which have taken place in their respective spheres.'[63]

Then there is Dick's jaw-dropping estimate of the Solar System population. Featured in his *Celestial Scenery* (1837), Dick discusses the cosmos, as seen from Mars, Jupiter, the planetoids, and beyond. Using a method similar to that of Huygens, Dick takes the topic further still, allotting populations to all the planetary bodies of the Solar System, and even for the rings of Saturn, a rather arresting idea.

Dick's logic is an impeccable example of cosmic imperialism. Beginning with an approximation of the population of Victorian England (Victoria had assumed the throne in the very year *Celestial Scenery* was published), Dick thus populates the heavens:

> The population of the different globes is estimated, as in the proceeding description, at the rate of 280 inhabitants to a square mile, which is the rate of population in England, and yet this country is by no means overstocked with inhabitants, but could contain, perhaps, double its

> present population. From the above statement, the real magnitude of all the moving bodies connected with the solar system may at once be perceived. If we wish to ascertain what proportion these magnitudes bear to the amplitude of our own globe, we have only to divide the different amounts stated at the bottom of the table by the area, solidity, or population of the Earth [see Table 4.1] ... the amount of population which the planets might contain [is] 21,894,974,404,480 ... Such is the immense magnitude of our planetary system, without taking into account either the Sun or the hundreds of comets which have been observed to traverse the planetary regions.[64]

Passing over the possibility of oceans elsewhere, Dick's table assigns to every planet and planetoid in our system, except Vesta, a population higher than Earth's. And, using today's number convention, he comes to the conclusion that there are roughly twenty-one trillion souls living in the Solar System alone.

Dick's was a crowded cosmos. His omission of a population estimate for the Sun does not mean the proselytising Dick was sceptical about Solarians – far from it. Using Herschel's comments and observations to support his claims, Dick announces, 'It would be presumptuous in man to affirm that the creator has not placed innumerable orders of sentient and intelligent beings ... throughout the expansive regions of the Sun.'[65]

Dr Thomas Dick was no fool. He knew enough astronomy to realise his opponents would have grave misgivings about some of his more extravagant claims. For instance, some eminent astronomers opposed Herschel's idea of populating luminous bodies such as the Sun. Others refused to accept the idea of life on planets that showed no evidence of an atmosphere. Dick had an answer for those too. Such planets had 'invisible' atmospheres that were 'purer' than ours, he claimed, and, 'the moral and physical condition of their inhabitants is probably superior to that enjoyed upon the Earth'.[66]

In the end, however, Dick was armed not with science but with faith. His major weapon was 'not telescopes but teleology',[67] as one commentator put it. But Dick was strident in his pluralism:

> Without taking pluralism into account, we can form no consistent views of the character of Omnipotence. Both his wisdom and his goodness might be called into question, and an idea of the Supreme Ruler presented altogether different from what is exhibited by the inspired writers in the records of Revelation.[68]

Dick was even to change his mind about life on comets. In 1840 he first published *Sidereal Heavens and Other Subjects Connected with Astronomy, As Illustrative of the*

Table 4.1. *The population of the Solar System, according to the work of Thomas Dick (1774–1857)*[121]

	Square Miles	Population	Solid Contents
Mercury	32,000,000	8,960,000,000	17,157,324,800
Venus	191,131,911	53,500,000,000	248,475,427,200
Mars	55,417,824	15,500,000,000	38,792,000,000
Vesta	229,000	64,000,000	10,035,000
Juno	6,380,000	1,786,000,000	1,515,250,000
Ceres	8,285,580	2,319,962,400	2,242,630,320
Pallas	14,000,000	4,000,000,000	4,900,000,000
Jupiter	24,884,000,000	6,967,520,000,000	368,283,200,000,000
Saturn	19,600,000,000	5,488,000,000,000	261,326,800,000,000
Saturn's Outer Ring	9,058,803,600	⇩⇩⇩⇩⇩⇩⇩⇩⇩	
Inner Ring	19,791,561,636	8,141,963,626,080	1,442,518,261,800
Edges of the Rings	228,077,000	⇧⇧⇧⇧⇧⇧⇧⇧⇧	
Uranus	3,848,460,000	1,077,568,800,000	22,437,804,620,000
The Moon	15,000,000	4,200,000,000	5,455,000,000
Jupiter's Satellites	95,000,000	26,673,000,000	45,693,970,126
Saturn's Satellites	197,920,800	55,417,824,000	98,960,400,000
Uranus' Satellites	169,646,400	47,500,992,000	84,823,200,000
AMOUNT	78,196,916,784	21,894,974,404,480	654,038,348,119,246

Character of the Deity and of an Infinity of Worlds. Here, if it can be believed, we find an even more evangelical Dick. Doing for the stars what *Celestial Scenery* did for the Solar System, *Sidereal Heavens* focuses on the sidereal system, but in passing through, Dick reconsiders comets in a chapter dubbed 'A Plurality of Worlds Proved from Divine Revelation'. His sceptical thoughts of *Christian Philosopher* (1823) are now transformed, as Dick declares comets not only habitable but claims, 'some of the comets ... may be peopled with intelligences of a higher order than the race of man'.[69]

The catalyst that sparked this change in Dick's attitude to comets seems to have been Johann Heinrich Lambert. Lambert, an eighteenth-century Swiss mathematician astronomer, was one of the original proponents of the nebular hypothesis of the origin of the Solar System, published in his 1761 treatise *Cosmological Letters* (*Cosmologische Briefe über die Einrichtung des Weltbaues*). Dick seems to have been fascinated by this knowledgeable Swiss Newtonian. Lambert incorporated into his system the idea that comets journey between star systems, and are inhabited by scientifically adept aliens. The comets, according to Lambert, can be expanded at specific times, preserving a habitable atmosphere.

Dick quotes copious amounts of Lambert. Witness this following passage from a book of Lambert's that William Herschel claimed was 'full of the most fantastic imaginations',[70] and in which an interlocutor describes the journey of these cometary alien travellers, from system to system:

> I rank highest the astronomers which you, Sir, place on such celestial bodies. Their route proceeds from suns as we go from city to city on Earth, and when in our case a few days go by they count in myriads of our years ... Our greatest measures are their differentials, and our millions can hardly suffice as their table of multiplication. They know the warmth and brightness of each sun, and with a single conclusion they determine the general characteristics of the inhabitants of each planet, which orbits around it at a given distance. Their year is the time from one sun to another. Their winter falls in the middle of the intervening space, or of the journey, which they make to another sun, and they celebrate the moment when their former course turns into a new one. The perihelion of each course is their summer.[71]

Like Dick, Lambert made all the heavenly bodies inhabitable, placing on each, 'innumerable inhabitants of all possible kind and form'.[72] And like Dick, Lambert held that 'The Creator ... is much too efficient not to imprint life, forces, and activity on each speck of dust ... If one is to form a correct notion of the world, one should set as a basis God's intention in its true extent to make the whole world inhabited.'[73]

In his *Sidereal Heavens*, Dick also considered alien appearance. In a chapter snappily entitled 'On the Physical and Moral State of the Beings That May Inhabit Other Worlds', he rejects the idea of the gargantuan alien of Voltaire and the like. Such mountainous size, Dick believed, would lead to overcrowding and be 'injurious to the exercise of intellectual faculties'.[74] Alien communication also crops up. To the speculation as to 'whether we may ever enjoy an intimate correspondence with beings belonging to other worlds',[75] Dick proposes that, though this is not presently possible, man is 'destined to a future and eternal state of existence, where the range of his faculties and his connexions with other beings will be indefinitely expanded'.[76]

The cosmologies of Herschel, Lambert, and Laplace lurk in the many passages of Dick's pluralist prose. And yet, as can be witnessed from the above account, his brand of pluralism is very much his own. Dick's imagined cosmic landscapes are the product of a fertile and esoteric religious consciousness, one that was determined to embrace the latest findings of cosmology, before it was too late, for the piously minded. But not each and every cosmic phenomenon is claimed for God. To the claim, for instance, by Samuel Vince, the Plumian

Professor of Astronomy at Cambridge, that the fading of a star is the 'destruction of that system at the time appointed by the Deity for the probation of its inhabitants',[77] Dick rebels, countering that, though God creates worlds, he does not destroy them.

The many worlds of Dr Thomas Dick were a significant milestone in the history of pluralism. And his influence was huge. Witness alone the Great Moon 'Hoax' with which we opened this chapter. Though the articles were a satire of Dick's work, they nonetheless recognised his far-reaching influence. Dick's swansong was his *Practical Astronomer*, a book published twelve years before his death in 1857. It contained little pluralism. Perhaps by now he had become less inured to the criticisms of his unique but esoteric blend of science and faith. And yet his habitually bizarre speculations on extraterrestrial life found favour with many. Among those notable travellers who journeyed all the way to Dundee to seek out Dick were a trio of notable Americans: essayist Ralph Waldo Emerson, author and abolitionist Harriet Beecher Stowe, and astronomer E. E. Barnard.

William Whewell: Giordano Bruno of the machine age?

At the top of this chapter we spoke of the direct effect of science on the Industrial Revolution, the machine age. But science also had an indirect effect, through its influence on the ruling ideas of the age. The revolution of thought in physics in the first half of the nineteenth century was not as profound as the seventeenth. In fact, it might be more accurate to call it reform, rather than revolution. For the nineteenth century saw a dissemination of those earlier revolutionary ideas into other fields of science, as articulated in the Newtonian synthesis.

As we also reported in the last chapter, when Atomism was resurrected, and the efforts of the Copernicans set in train, the Christian picture of creation suffered a near mortal blow. But Newton's system of the world had essentially re-established the integrity of design, which had been shattered by the Copernicans. The work of the Caledonian canons, Thomases Chalmers and Dick, had been to claim for Christians the wonders of the modern universe, removing the obstacles in the way of belief. The Christian picture of Creation had survived. Man was still made in the image of God. But there was worse to come. Darwin was to strike at the heart of humanity itself. After Darwin, the book of Genesis lay in tatters as a literal history.

So the imminent revolution was to be in biology. The revolution in this rather descriptive science would have happened long before, but for the opposition from landed and clerical interests. Intuitively they knew that the bottom-up approach of materialist science once more threatened any rationalisation of a divine ordering of the world. As with the seventeenth-century revolution in physics, it would be some time before an appropriate face-saving theistic formula could be found for the Darwinian revolution. A lull in which they scrambled to discover religious truth in some remote way that could not be contradicted by vulgar facts.

Perhaps in anticipation of such future battles, a single book, William Whewell's *Of the Plurality of Worlds: An Essay*, dominated the mid-nineteenth-century pluralist debate. It was a period of intense debate in the question of extraterrestrial life. The book was published in 1853, amid an outcry when Whewell declared many pluralist arguments to be scientifically inept, and religiously treacherous. In the history of pluralism, Whewell's book certainly warrants further examination.

William Whewell was an English polymath, perhaps best known for his coining of the words 'scientist' and 'physicist'. Known variously as philosopher, theologian, and historian of science, Whewell had been born in Lancaster, the son of a master carpenter. Rather than follow the old man in his trade, Whewell had won an exhibition to study at Trinity College, Cambridge. Years later, he was to become Master of the College. Awarded the Chancellor's Gold Medal for poetry in 1814 while still a student, Whewell rose through the ranks, becoming President of the Cambridge Union in 1817, then a fellow and tutor of his College. In 1841, he succeeded Dr Christopher Wordsworth, younger brother of the poet William, as Master, holding the chairs of mineralogy from 1828 to 1832, and philosophy from 1838 to 1855.

As Darwin had begun his career a creationist, Whewell had started his a pluralist. Around 1820, Whewell had discovered the debate about extraterrestrial life. Drawn first to the works of Fontenelle, Whewell then came upon Thomas Chalmers. In 1822, Whewell wrote to his sister, 'I am glad that you have been reading more of Chalmers' sermons; I have read only one or two of the astronomical ones, but I think they are all that you describe them.'[78] Five years later, and one year after his ordination in the Anglican Church, Whewell gave a series of four sermons in the University Church at Cambridge. The third sermon clearly shows his eagerness for the pluralist message of Chalmers:

> the Earth ... is one among a multitude of worlds ... with resemblances and subordinations among them suggesting ... that [they may be inhabited by] crowds of sentient ... beings – these are not the reveries of idle dreamers or busy contrivers; but, for the main part, truths collected by wise and patient men on evidence indisputable, from unwearied observation and thought; and for the rest, founded upon analogies which most will allow to reach at least to some degree of probability.[79]

Whewell was an idealist. By the time he had been elected a fellow of Trinity College, he had pored over *Critique of Pure Reason*, by German philosopher Immanuel Kant. Kant too had been a pluralist of the idealist kind. In an effort to counter the atheistic implications of Epicureanism, he had opposed those who 'assign to a nature left to itself such processes in which one rightly perceived the immediate hand of the Supreme Being'.[80] Nonetheless, Kant was convinced of the probability of planetary life as 'it is not even necessary to assert that all planets must be inhabited, although it would be sheer madness to deny this in respect to all, or even to most of them'.[81] Kant's influence was to provide Whewell with the philosophical position to oppose the dominant empiricism of the day.

Whewell's book caused a great stir when it was published in 1853. For one thing, it was very much against the pluralist stream. For another, the challenger was none other than William Whewell. A review of note was printed in London's *Daily News*, founded in 1846 by Charles Dickens, who also served as the paper's first editor. This is what the paper had to say about Whewell's *Of the Plurality of Worlds*:

> We scarcely expected that in the middle of the nineteenth century, a serious attempt would have been made to restore the exploded idea of man's supremacy over all other creatures in the universe; and still less that such an attempt would have been made by one whose mind was stored with scientific truths. Nevertheless a champion has actually appeared, who boldly dares to combat against all the rational inhabitants of other spheres; and though as yet he wears his vizor down, his dominant bearing, and the peculiar dexterity and power with which he wields his arms, indicate that this knight-errant of nursery notions can be no other than the Master of Trinity College, Cambridge.[82]

In the furor that followed, Whewell's book went through more than five English and two US editions,[83] and bred more than seventy published rejoinders. Many of these responses were themselves books, which in turn inspired Whewell to write and publish an answer to his critics, in the form of his anonymous 1854 book, *Dialogue on the Plurality of Worlds*.[84] Since Whewell's books on pluralism are replete with original arguments, some of which are still relevant today, they certainly bear a detailed examination. Indeed, they serve as an ideal spotlight for the debate on pluralism that raged in the middle of the nineteenth century, inspired by Whewell himself.

The preface to *Of the Plurality of Worlds* is witness to Whewell's fear of becoming a Giordano Bruno of the machine age. He writes, 'It will be a curious, but not very wonderful event, if it should now be deemed as blameable to doubt the existence of inhabitants of the Planets and Stars as, three centuries ago, it was held heretical to teach that doctrine.'[85] A full three centuries would take us back to

Copernicus, of course, but since Copernicus was a rather reluctant Copernican, Whewell must have been referring to the braver Bruno.

It was quite a serious issue for Whewell, this prospect of ridicule and sacrilege. In a letter to J. D. Forbes in November of 1853, he states his fears that his book might be considered heretical. Later he points out that, although Christianity for most of its long history had no need for pluralist doctrine, nonetheless, 'at the present day ... many persons have so mingled this assumption with their religious belief, that they regard it as an essential part of Natural Religion'.[86] Whewell then signs off his introduction with the thought that some readers may find pluralism to be at odds with their faith. His book may provide succour to those unhappy with the fashionable blend of religion and science. For some pious Victorians, idealism laced with materialism was not a pleasing mixture.

On the plurality of worlds

The first chapter of Whewell's book gazes upon the heavens through the lens of the psalmist. Like Chalmers before him, Whewell quotes from the Eighth Psalm, 'When I consider the heavens, the work of thy fingers, the moon and the stars, which thou has ordained; what is man, that thou art mindful of him? And the son of man, that thou visitest him?' In Whewell's view, the psalmist was not sceptical of God as a watcher of the skies. Rather, he is simply asking, '*What* is man?', that he should be so set apart from the cosmos. The question was a simpler one under Aristotle's geocentric sky. But for the nineteenth-century reader, blessed with the benefit of Herschel throwing open the deep sky to closer observation, the answers are not so prosaic. Whewell then lists the findings of modern astronomy, this opening chapter being titled 'Astronomical Discoveries'. He sets the pluralist case: the planets may be other Earths, the stars other Suns, also orbited by other worlds, and the Milky Way a seething mass of innumerable suns.

In Chapter 2, 'The Astronomical Objection to Religion', Whewell focuses on the argument whose distilled essence he quotes in this way:

> if this world be merely one of innumerable worlds ... occupied by intelligent creatures, [then] to hold that this world has been the scene of God's care and kindness, and still more, of his special interpositions, communications, and personal dealings with its individual inhabitants ... is, the objector is conceived to maintain, extravagant and incredible.[87]

In Whewell's skilful hands, this argument is not considered 'as an objection, urged by an opponent of religion, but rather as a difficulty, felt by a friend of religion'.[88] Faced with modern astronomy, Whewell sounds as if he would prefer the clock be turned back to the pre-Copernican age of faith: 'How shall the earth, and men, its inhabitants, thus repeatedly annihilated, as it were by the growing magnitude of the known universe, continue to be anything in the regard of Him who embraces all?'[89] One eminent commentator has pointed out the Pascalian nature of this passage of Whewell's.[90] A seventeenth-century French polymath, Blaise Pascal had been a sceptic on the question of life elsewhere. Gifted with scientific flair and a religious sensibility, Pascal appears to have pondered the problem of pluralism in his *Pensées*, his most influential theological work. The books contains somewhat mysterious and compelling remarks, such as, 'The eternal silence of these infinite spaces frightens me'.[91] American historian Arthur Lovejoy was of the opinion that in Pascal we find 'the curious combination of a refusal to accept the Copernican hypothesis with the unequivocal assertion of the Brunonian'.[92] Witness the fact that, in his ordering of the *Pensées*, Pascal's 'infinite spaces' sentence is tagged by the remark, 'How many kingdoms know us not!'[93] Lovejoy concludes, 'Pascal seems to conceive of mankind as alone in a dead infinity of matter'.[94]

Chapter 3 of Whewell's book is entitled 'The Answer from the Microscope'. Here the question of microscope and telescope comes under his hammer. Chalmers had said that those who doubt God's custody of man in a cosmos so vast should contemplate the microscope, whose discoveries implied divine care even of creatures so small. Though praising Chalmers, Whewell begs to differ. The problem for the pious, he suggests, originates in the anxiety created not from pluralism per se, but from belief in *intelligent* life. The microscope is no help, says Whewell, not unless tiny intelligent creatures can be spied beneath.

In Chapter 4, Whewell wonders about the culture of alien life. What if, he asks, other worlds are peopled by 'creatures analogous to man; – intellectual creatures, living . . . under moral law, responsible for transgression, the subjects of a Providential Government'.[95] Whewell then contemplates the cultural history of such an alien race, 'Wherever pure intellect is, we are compelled to conceive that, when employed upon the same objects, its results and conclusions are the same.'[96] A rather sweeping and geocentric conclusion, he implies. But from this premise, Whewell suggests, we conclude that any intelligent aliens will have 'the intellectual history of the human species . . . They must have had their Pythagoras, their Plato, their Kepler, their Galileo, their Newton.'[97]

Based upon such views of extraterrestrials from Fontenelle on, Whewell says, we tend to think of intelligent creatures in the cosmos as effectively identical to us. Even down to a similar history and identity. Witness, says Whewell, that 'the great astronomer Bessel had reason to say that those who imagined inhabitants in the Moon and Planets, supposed them, in spite of all their protestations, as like to men as one egg to another'.[98] In this way, Whewell reduces the argument to its geocentric absurdity. To suggest that aliens have a nature and history even similar to man's is 'an act of invention and imagination which may be as coherent as a fairy tale, but which, without further proof, must be as purely imaginary and arbitrary'.[99]

Whewell uses the methods of science, but in defence of faith. However, having railed against geocentrism, he then falls for the same conceit in arguing a special case for Earth and man. God, he says, has sent a special message to the Earth:

> The arrival of this especial message ... forms the great event in the history of the Earth ... It was attended with the sufferings and cruel death of the Divine Messenger thus sent; was preceded by prophetic announcements of his coming; and the history of the world, for the two thousand years that have since elapsed, has been in a great measure occupied with the consequences of that advent. Such a proceeding shows, of course, that God has an especial care for the race of man.[100]

His conclusion was to prove controversial:

> The Earth ... can not, in the eyes of any one who accepts this Christian faith, be regarded as being on a level with any other domiciles. It is the Stage of the great Drama of God's Mercy and Man's Salvation ... This being the character which has thus been conferred upon it, how can we assent to the assertions of Astronomers, when they tell us that it is only one among millions of similar habitations ... ?[101]

Next, Whewell embarks on an anti-pluralist cosmic journey. Running through chapter headings such as 'Geology', 'The Fixed Stars', 'The Planets', and 'The Theory of the Solar System', he sets out his scientific thesis against the idea of extraterrestrial life. His first stop is the home planet, and Earth history. Whewell responds to the question of why man should make such a late appearance on Earth, if he was the one crucial part of God's plan. Rather than this being a colossal waste of time, it is indicative of the way God works. Just as human history is but an 'atom of time', we should not doubt that intelligent life is restricted to a mere 'atom of space'.

When he turns to the stars, Whewell takes issue with the notion that they are in nature just like our Sun. He wonders whether the objects espied in star clusters are actually stars at all. This scepticism can be partly forgiven when we consider that comparatively little stellar data was actually known at the time Whewell was writing. Hence he could argue that 'so far as we yet know, the Sun is the largest sun among the Stars'.[102] Elsewhere, Whewell was thoughtful on the variety of stellar possibilities: double star systems may not prove suitable for planets; other single stars may be orbited by planets, but there is no evidence;

and variable stars are not truly like our Sun anyhow. And so he was able to contend that the assumed equivalence of our Sun and the stars was pitched on a slippery slope. Whewell quotes German naturalist Alexander von Humboldt, who had wondered, 'is the assumption of satellites to the fixed stars so absolutely necessary? If we were to begin from the outer planets, Jupiter, &c., analogy might seem to require that all planets have satellites. But yet this is not true for Mars, Venus, Mercury.'[103]

Whewell's conclusion was simple. Readers should shy away from speculation about aliens on other planets until evidence of such planets is actually found. One wonders, if Whewell had been alive in 1995, whether he would have allowed such speculation after the discovery of exoplanets. Having attempted to drain the cosmos of all imaginings, he then turns his attentions to the Solar System, beginning with the recently discovered Neptune. Cold and remote, and lacking in light and heat, he considers Neptune ill-suited to life, unless a rule can be found by which planets are peopled. Our Moon is another matter. Armed with the research of German astronomer Friedrich Bessel, Whewell challenges the notion that the Moon is favoured for life. Lacking both water and an atmosphere, he concludes the Moon to be lifeless, and conjecture about life on the planets pointless.

Whewell moves from planet to planet, dismissing each in turn. Jupiter, with its lower density but huge mass is considered 'a mere sphere of water', which, if it harbours any life at all, it would be 'cartilaginous and glutinous masses, boneless, watery, pulpy creatures', suspended in solution.[104] Saturn fares no better. Devoid of light and heat, and less dense even than Jupiter, it is considered an even less likely habitat for intelligent aliens. After dispensing with Uranus and Neptune with similar contempt, Whewell then turns his attentions to Mars.

Mars was not so easy to dismiss. Resembling the Earth in terms of continents and ice caps, seas and night and day, Mars nonetheless receives less light and heat than our home planet, and is half the size. Bearing this in mind, Whewell thinks of Mars as an early Earth, populated by dinosaurs and the like. Skipping quickly over the asteroids, he then makes his way to Venus and Mercury, whose proximity to the Sun renders them barren, though Whewell does allow the possibility of Venusian 'microscopic creatures, with siliceous coverings'.[105]

Whewell goes all Goldilocks

And so Whewell's cosmic odyssey is complete. Having taken his readers on a magical mystery tour from the outer reaches of the universe back down to Earth, he makes a robust case against pluralism. To be fair, his opponents had made his ride a rather easy one. Their compulsion to speculate wildly, on the minimum of available evidence, had gifted Whewell something of an open goal. A man of considerable intellect, he understood how weak was the astronomy, how scant the science. And Whewell was right about the absence of intelligent life in the Solar System, even though the debate about Mars was to rage on, late into the century and beyond. Whewell's *Of the Plurality of Worlds* was a significant milestone in the astronomical debate on alien life.

In a short Chapter 10, Whewell attempts a synthesis. His intention is to squeeze his hypotheses from earlier chapters into a theory of the Solar System. His inner system is a region of hot, solid bodies; the outer system a cold region of giant planets of viscous fluidity. Only the Earth occupies an orbit that is 'the Temperate Zone of the Solar System'.[106] In Whewell's schema, God has seen to it that his planet-forming powers are greatest near the Earth, becoming weaker further out:

> The Earth is really the domestic hearth of this Solar System; adjusted between the hot and fiery haze on one side, the cold and watery vapour on the other. This region only is fit to be a domestic hearth, a seat of habitation; and in this region is placed the largest solid globe of our system; and on this globe, by a series of creative operations, entirely different from any of those which separated the solid from the vaporous, the cold from the hot, the moist from the dry, have been established, in succession, plants, and animals, and man ... the Earth alone ... has become a World.[107]

Aware of the fact that his theories are novel and challenging, Whewell urges the reader to consider his arguments with a careful but open mind. He admits that in some respects his own creation and the nebular hypothesis bear some resemblance:

> The Nebular Hypothesis ... is that part of our Hypothesis, which relates to the condensation of luminous nebular matter; while we consider, further, the causes which, scorching the inner planets, and driving the vapours to the outer orbs, would make the region of the Earth the only habitable part of the system.[108]

Finally, in the last three chapters, Whewell delivers his concluding thoughts. Focusing on the cultural milieu in which pluralism has gained ground, he examines the religious and scientific reasons for the notion of extraterrestrial life proving so popular. The first of these is a chapter entitled 'Arguments from Design'. Here, Whewell's aim is to patch up his anti-pluralist stance with theology. The teleological argument, he suggests, is fair as well as foul. Consider the wonderful adaptations in nature. The limb of a human, or the wing of a bird.

Each has its own special function, and yet they are conspicuously similar in form. Whewell offers that such inconsistencies can be removed if we assume God plays with certain patterns, adapted to specific conditions. That the wing and the limb may be created from the same blueprint does not destroy the argument from design:

> If the general law supplies the elements, still a special adaptation is needed to make the elements serve such a purpose; and what is this adaptation, but design? The radius and ulna, the carpal and metacarpal, are all in the general type of the vertebrate skeleton. But does this fact make it less wonderful, that man's arm and hand and fingers should be constructed so that he can make and use the spade, the plough, . . . the lute, the telescope . . . ?[109]

Indeed, suggests Whewell, the presence of a blueprinted plan may indicate a faculty of what he refers to as 'the Creative Mind'. There may not be a cosmos replete with life, but God's plan may instead mean this: that his pattern of creation forms planets and stars aplenty, the notable product of which is our single habitable world. God did not create those other worlds in vain. Rather, they are merely a part of the pattern, seeds that did not flower: 'Of the vegetable seeds which are produced, what an infinitely small portion ever grow into plants! Of animal ova, how exceedingly few become animals, in proportion to those that do not; and that are wasted, if this be waste!'[110]

And so, in the hands of The Great Gardener, our Earth becomes the only 'fertile seed of creation', or the 'one fertile flower [of] the Solar System . . . One such fertile result as the Earth, with all its hosts of plants and animals, and especially with Man . . . is worthy and sufficient produce . . . of all the Universal Scheme.'[111] Readers should not trouble themselves that vast swathes of the cosmos are left without intelligent life. After all, such was the Earth for most of its history.

In this way, Whewell attempts an idealist's reconciliation. On the one hand we have the yawning splendour of the vast and modern cosmos, with all its possibilities and potential. On the other we have Whewell, desperately attempting to square this materialist vision with his own thoughts on man as the sole sentient creature in the cosmos. He does so with considerable colour:

> The planets and the stars are the lumps which have flown from the potter's wheel of the Great Worker; – the shred-coils which . . . sprang from His mighty lathe: – the sparks which darted from His awful anvil when the solar system lay incandescent thereon; – the curls of vapour which rose from the great cauldron of creation when the elements were separated. If even these superfluous portions are marked with universal

> traces of regularity and order, this shows that universal rules are his implements, and that Order is the first and universal Law of the heavenly work.[112]

Whewell as geocentric champion

In the final analysis, however, and with geocentric aplomb, Whewell declares the universe as nothing, compared with man. Indeed, it seems that just 'one soul created never to die ... outweighs the whole unintelligent creation'.[113] Religion, he suggests, has little to lose by giving up pluralism:

> The majesty of God does not reside in planets and stars ... which are ... only stone and vapour, materials and means ... the material world must be put in an inferior place, compared with the world of mind. If there be a World of Mind, that ... must have been better worth creating ... than thousands and millions of stars and planets, even if they were occupied by a myriad times as many species of brute animals as have lived upon the Earth since its vivification.[114]

With his book drawing to a close, Whewell comes clean about his Platonism. His idealism is laid bare with his notion of the 'Divine Mind', which man can come to know through the study of nature and its archetypes. Such knowledge sets man apart, he says. Man's intellect gains a measure of harmony with the Divine Mind, which justifies Whewell's idea, 'why, even if the Earth alone be the habitation of intelligent beings, still the great work of creation is not wasted'.[115] Finally finding the fix he needs, Whewell declares that the 'remotest planet is not devoid of life for God lives there'.[116] God's presence is down to His laws of the universe. And given that He is in all cosmic places, Whewell is sceptical that 'the dignity of the Moon would be greatly augmented if her surface were ascertained to be abundantly peopled with lizards'.[117]

Indeed, Whewell's inflated geocentrism now knows no bounds. Because man can attain divine moral law, he is ample validation for the whole cosmos: 'One school of moral discipline, one theatre of moral action, one arena of moral contests for the highest prizes, is a sufficient centre for innumerable hosts of stars and planets, globes of fire and earth, water and air, whether or not tenanted by corals and madrepores, fishes and creeping things.'[118] Thus is man. Not merely gifted the chance of moral eminence, man's ordeal on Earth is so profound it can result in everlasting life with God:

> The Interposition of God, in the history of man ... is an event entirely out of the range of those natural courses of events which belong to our subject [pluralism]; and to such an Interposition, therefore, we must

> refer with great reserve . . . But this, it would seem, we may say; – that such a Divine Interposition . . . is far more suitable to the Idea of God of Infinite Goodness, Purity, and Greatness, than any supposed multiplication of a population . . . not provided with such means of moral and spiritual progress.[119]

Finally, Whewell looks to 'The Future' in his last chapter. What will become of man? What does the future hold? Whewell is acutely aware of a loophole in his arguments on geology, developed in earlier chapters. If human history is but an 'atom of time', the reader might ask, why might man not become extinct? After all, the geologists of the machine age were busy opening up the veins of the Earth, digging up signatures of beasts no longer found anywhere on the globe. Might man meet the same fate?

Man is the exception, Whewell predictably concludes. Indeed, the emergence of man is the great event of our geological past. And even though human history occupies but a second of time, man will endure. No reader pondering creation would have wagered on the emergence of man. And his progress has been swift and compelling. But as to the machine age, Whewell is sceptical it will truly advance man any further, 'Men might be able to dart from place to place, and even from planet to planet, and from star to star, on wings, such as we ascribe to angels in our imagination: they might not be able to make the elements obey them at a beck: yet they might not be better, or even wiser, than they are.'[120]

In a curiously prescient comment, with respect to science fiction narratives of the near future, Whewell was at pains to deny the possibility of a super man – Man is indeed superior to the brute. But Whewell foresees no coming of a race superior to man. And yet, within the decade, Darwin's *Origin of Species* was to infuse science with an even greater sense of history. And the theory of evolution would inspire an explosion in futuristic fantasies, in which the alien was to the fore.

Notes

1 Leibniz, G. W. (1710) *Théodicée*, trans. E. M. Huggard (1951) London
2 Carlyle, T. (1904) *Signs of the Times* (in *Critical and Miscellaneous Essays*), New York
3 Quoted in Locke, R. A. (1859) *The Moon Hoax, or A Discovery that the Moon Has a Vast Population of Human Beings*, New York, p. 61
4 ibid., p. 78
5 ibid., p. 79
6 ibid., pp. 95–6
7 ibid., p. 96
8 ibid., p. 98
9 ibid., p. 105
10 ibid., p. 109

11 ibid., p. 113
12 ibid., pp. 115–16
13 ibid., p. 61
14 ibid., p. 61
15 ibid., p. 62
16 ibid., p. 62
17 Moss, S. P. (1963) *Poe's Literary Battles*, Durham, p. 87
18 Anonymous (1853) Locke Among the Moonlings, *Southern Quarterly Review*, **24**, if the account is to be taken as fact, or fact laced with satire
19 ibid., p. 502
20 Poe, E. A. (1902) Richard Adams Locke, in *Complete Works of Edgar Allan Poe*, vol. XV, ed. James A. Harrison, New York, pp. 126–37
21 O'Brien, F. M. (1928) *The Story of the Sun*, New York, pp. 37–57
22 Poe, E. A. (1902) Richard Adams Locke, in *Complete Works of Edgar Allan Poe*, vol. XV, ed. James A. Harrison, New York p. 136
23 Crowe, M. J. (2008) *The Extraterrestrial Life Debate, Antiquity to 1915, A Source Book*, Notre Dame
24 ibid., p. 275
25 Crowe, M. J. (1986) *The Extraterrestrial Life Debate, 1750–1900*, New York, p. 215
26 Griggs, W. N. (1852) *The Celebrated 'Moon Story', Its Origin and Incidents with a Memoir of Its Author*, New York, pp. 8–9
27 ibid., p. 8
28 Known as kerosene in the United States, Canada, and other parts of the world
29 Griggs, W. N. (1852) *The Celebrated 'Moon Story', Its Origin and Incidents with a Memoir of Its Author*, New York, p. 33
30 ibid.
31 ibid.
32 Anonymous (1853) Locke Among the Moonlings, *Southern Quarterly Review*, **24**, 502
33 Quoted in Crowe, M. J. (2008) *The Extraterrestrial Life Debate, Antiquity to 1915, A Source Book*, Notre Dame, p. 240
34 Warren, S. (1854) Speculators among the Stars, *Blackwood's Edinburgh Magazine*, **76**, 373
35 Hanna, W. (1849) *Memoirs of the Life and Writings of Thomas Chalmers*, 4 vols, Edinburgh, 2:89
36 As quoted in Hanna, *Memoirs*, 2:89
37 Hitchcock, E. (1854) Introductory Notice to the American Edition of William Whewell's *The Plurality of Worlds*, Boston
38 Watt, H. (1943) *Thomas Chalmers and the Disruption*, London, p. 51
39 Chalmers, T. (1817) *A Series of Discourses on the Christian Revelation Viewed in Connection with the Modern Astronomy*, Edinburgh, pp. 3–4
40 ibid.
41 ibid., p. 24
42 ibid., p. 30
43 ibid., p. 66
44 ibid., p. 73
45 ibid., p. 100
46 ibid., p. 107
47 ibid., p. 109

48 ibid., p. 118
49 ibid., p. 124
50 Bode, J. E. (1778) Gedanken über die Natur der Sonne und Entstehung ihrer Flekken, *Beschäftigungen der Berlinischen Gesellschaft Naturforschender Freunde*, 2, Berlin, p. 246
51 Cairns, D. (1973) Thomas Chalmers's *Astronomical Discourses: A Study in Natural Theology*, in *Science and Religious Belief*, ed. C.A. Russell, Bungay, UK, pp. 195–204
52 Dick, T. (1846) *Celestial Scenery, or the Wonders of the Planetary System Displayed; Illustrating the Perfections of Deity and a Plurality of Worlds*, Hartford, pp. 161, 163, 166, and 171–2 (first edition 1837)
53 Dick, T. (1848) *Sidereal Heavens and Other Subjects Connected with Astronomy, As Illustrative of the Character of the Deity and of an Infinity of Worlds*, Hartford, pp. 118, 122, and 123
54 Dick, T. (1848) *The Christian Philosopher, or the Connexion of Science and Philosophy with Religion*, Hartford, p. 30
55 ibid., p. 31
56 ibid., p. 81
57 ibid., p. 84
58 ibid., p. 143
59 Dick, T. (1844) *Philosophy of Religion* in *Works*, Hartford, p. 65
60 ibid., p. 68
61 Dick, T. (1828) *Philosophy of a Future State* in *Works*, Hartford, p. 103
62 ibid.
63 ibid.
64 Dick, T. (1846) *Celestial Scenery* in *Works*, Hartford, p. 135–6
65 ibid.
66 ibid., p. 168
67 Crowe, M. J. (1986) *The Extraterrestrial Life Debate, 1750–1900*, New York, p. 200
68 Dick, T. (1848) *Celestial Scenery* in *Works*, Hartford, p. 176
69 Dick, T. (1848) *Sidereal Heavens and Other Subjects Connected with Astronomy, As Illustrative of the Character of the Deity and of an Infinity of Worlds*, Hartford, p. 168
70 Herschel, W. (1912) *On the Direction and Velocity of the Motion of the Sun, and Solar System* in *Scientific Papers*, ed. J. L. E. Dreyer, vol. II, London, p. 318
71 Lambert, J. H. (1976) *Cosmological Letters on the Arrangement of the World-Edifice*, trans. S. L. Jaki, p. 111. The original is Lambert (1761) *Cosmologische Briefe über die Einrichtung des Weltbaues*, Augsburg
72 ibid., p. 55
73 ibid., p. 102
74 Dick, T. (1848) *Sidereal Heavens and Other Subjects Connected with Astronomy, As Illustrative of the Character of the Deity and of an Infinity of Worlds*, Hartford, p. 135
75 ibid., p. 139
76 ibid.
77 ibid., p. 47
78 Douglas, Mrs Stair (1882) *The Life of William Whewell*, 2nd edn., London, p. 74
79 As quoted in Todhunter, I. (1876) *William Whewell*, 2 vols, London, p. 326
80 Kant, I. (1981) *Universal Natural History and Theory of the Heavens*, trans. S. L. Jaki, Edinburgh, p. 81
81 ibid., p. 184
82 As quoted in Samuel Warren (October 1854) Speculators among the Stars, Part II, *Blackwood's Edinburgh Magazine*, **76**, p. 372

83 Whewell, W. (1853) *Of the Plurality of Worlds: An Essay*. The English editions appeared in 1853, summer 1854, autumn 1854, 1855, and 1859. The US editions came out in 1854 and 1861. References in the main body of the text are to the London 1853 first edition.
84 Whewell, W. (1853) *Dialogue on the Plurality of Worlds, Being a Supplement to the Essay on That Subject* London. This work was reprinted in the *Essay* in later editions.
85 Whewell, W. (1853) *Of the Plurality of Worlds: An Essay*, London, p. iii
86 ibid., pp. iii–iv
87 ibid., p. 19, as quoted in M. J. Crowe (1986) *The Extraterrestrial Life Debate, 1750–1900*, New York, p. 283
88 ibid., p. 20
89 ibid., p. 21
90 Crowe, M. J. (1986) *The Extraterrestrial Life Debate, 1750–1900*, New York, p. 283
91 Pascal's *Pensées*, intro. T. S. Eliot (1958) New York, p. 206
92 Lovejoy, A. (1936) *The Great Chain of Being*, Cambridge, Mass., p. 126
93 Pascal's *Pensées*, intro. T. S. Eliot (1958) New York, p. 207
94 Lovejoy, A. (1936) *The Great Chain of Being*, Cambridge, Mass., p. 129
95 Whewell, W. (1853) *Of the Plurality of Worlds: An Essay*, London, p. 30
96 ibid., pp. 31–2
97 ibid., pp. 36–7
98 ibid., p. 43
99 ibid., p. 37
100 ibid., p. 44
101 ibid., pp. 44–5
102 ibid., p. 160
103 As quoted in Whewell, *Essay*, pp. 161–2
104 ibid., pp. 179–83
105 ibid., p. 192
106 ibid., p. 196
107 ibid., p. 203
108 ibid., p. 208
109 ibid., p. 214
110 ibid., p. 222
111 ibid., p. 244
112 ibid., p. 243
113 ibid., p. 244
114 ibid.
115 ibid., p. 253
116 ibid., p. 252
117 ibid.
118 ibid., p. 256
119 ibid., p. 258
120 ibid., p. 274
121 Taken from Crowe, M. J. (2008) *The Extraterrestrial Life Debate, Antiquity to 1915, A Source Book*, Notre Dame, p. 270

5

After Darwin: *The War of the Worlds*

Two years have fled, O Lumen, since the day of our first mystical conversation. During this period, of which inhabitants of eternal space like you have been unconscious, but of which we dwellers on the Earth have been very conscious, I have often devoted my thoughts to the great mysteries into which you have initiated me, and to the new horizons set out before my mind's eye. Since your departure from the Earth you have doubtless made great advances, by means of your observations and studies, in progressively vaster fields of research. Doubtless, too, you have numberless marvels to relate to me now that my intelligence is better prepared to receive them. If I am worthy, and can comprehend them, please give me an account of the celestial voyages that have transported your spirit into the higher spheres, of the hitherto-unknown truths that they have imparted to you, of the grandeurs that they have exposed to you, and of the principles they have taught you in reference to the mysterious subject of the destiny of human and other beings.

Camille Flammarion, *Lumen*, trans. B. Stableford[1]

No one would have believed in the last years of the nineteenth century that this world was being watched keenly and closely by intelligences greater than man's and yet as mortal as his own ... With infinite complacency men went to and fro over this globe about their little affairs, serene in their assurance of their empire over matter ... At most, terrestrial men fancied there might be other men upon Mars, perhaps inferior to themselves and ready to welcome a missionary enterprise. Yet across the gulf of space, minds that are to our minds as ours are to the beasts that perish, intellects vast and cool and unsympathetic, regarded this Earth with envious eyes, and slowly and surely drew their plans against us.

H. G. Wells, *The War of the Worlds*[2]

The Great Chain of Being

Astrobiology, of course, has its roots in biology, as well as physics. For much of our narrative we have focused on progress in physics. The reason for this is clear when we take even a cursory look at history. As we mentioned earlier, writers and scholars considering the possibility of alien life were faced with a sequence in which the fields of experience were brought within the scope of science. Roughly it ran: mathematics, astronomy, physics, chemistry, and biology.

The reader will recall that an analysis of this evolution is not a trivial one, but that the order of the development seems to have been influenced by the interests of ruling or rising classes at significant times. An obsession with the calendar, a priestly occupation, inspired an early necessity for astronomy, and yet it was two millennia before the demands of the rising manufacturers of the seventeenth century spawned modern chemistry. Biology was another matter. The field was not prised open without very direct study. Even so, ancient ideas on nature and evolution either helped or hindered the progress of thoughts on life in the universe.

A kind of claustrophobic continuity was dominant. Like the walled-in towns of the medieval world, such was the medieval universe. A walled-in cosmos was the main currency, a space bounded between heavenly permanence and diabolical earthly change. But this was very much a *connected* cosmos. It was a creation that emanated from God in the highest realm, through the nested planetary spheres of crystalline perfection, down to the dark and lowly Earth at the corruptible centre.[3]

The system of Aristotle had been transformed, from a two-tier affair into an ornate pageant of divine creation. The simple split of the cosmos into sub-lunary and supra-lunary spheres had been preserved. Within its broad boundaries, though, now sat a cornucopia of creation, an endless pageant of links, extending from God down to the lowliest form of life. Such was the genesis of the Great Chain of Being.

The contrasting essence of Aristotle's original system had also been conserved. Permanence and change. This scale of being, the *scala naturae*, was an exacting hierarchy. It ranged from Godly perfection, at the top of the chain, down to the fallibility of flesh at the bottom, mutable and corruptible man. It was an order of nature. Those elements that sat at the foot of the chain, such as rock, merely existed. But on moving up the chain, each successive link enjoyed better attributes than those below. Plants possessed life, as well as existence; but animals were blessed with the qualities of motion and appetite.

All creatures in creation had a rung on this great ladder. Each place was decided according to its utility to man. Thus, in a somewhat anthropocentric

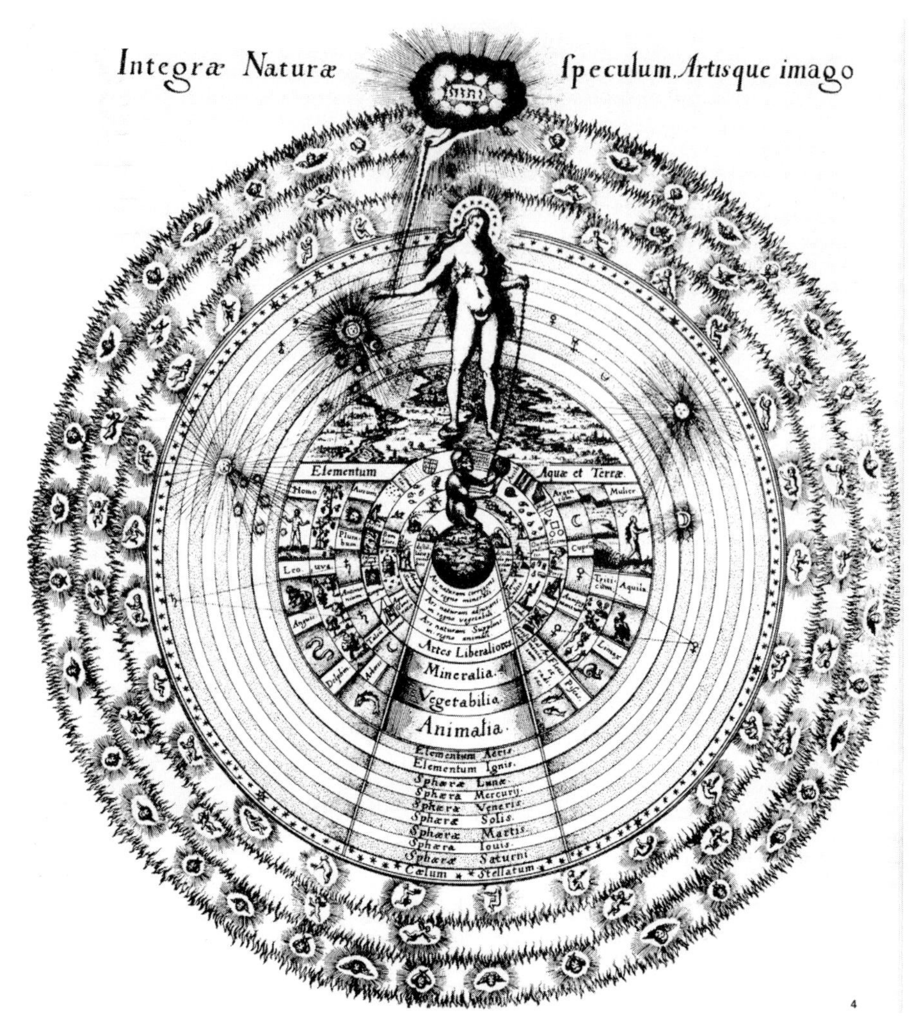

Figure 11 1579 depiction of the Great Chain of Being from Didacus Valades, *Rhetorica Christiana*; all of creation is here from the most basic and foundational elements, such as dirt, up to the very highest perfection, God.

way, wild beasts were superior to domestic ones, as they had refused training. Functional beasts, such as horses and dogs, were superior to docile ones, such as sheep. Paradoxically, readily trained birds of prey were better than lowlier birds, such as pigeons. And edible fish were higher up the ladder than the dubiously inedible creatures of the sea.

A sense of heavenly aesthetics was also injected into the *scala naturae*. Pleasing creatures, the likes of dragonflies and ladybirds, were thought better envoys of God's glory than unpleasant insects, such as flies and, probably, dung beetles.

The serpent unjustly languished at the very base of the animal section, demoted as retribution for its actions in the Garden of Eden. A few aspects of the Chain persist today in popular culture: the lion is yet regarded the king of all creatures, the oak tree king of plants.

The *scala naturae* was a hierarchy fixed in time.

It was not merely unthinkable to leave your place on the Great Chain; it was impossible. This austere permanence was key to the idealist conception of the world. And it is little surprise to see how the *scala* became a crucial means of control in feudal society. The Chain was used to validate the dogma of the Divine Right of Kings, the notion that the monarch was subject to no earthly authority. The *scala* had its echo in an authoritarian social order that saw kings at the summit, aristocratic lords beneath, and the great mass of peasants far down the social ladder. The king derived his right to rule from God. He was not subject to the will of his people, the aristocracy, or any other estate of the realm.

Man was a special case. He was mortal in flesh, but pure in spirit. The dialectic of this opposition meant that man could either go the noble way of the spirit, or be dragged down to the devil, the way of all flesh, just as Lucifer himself had fallen. The *scala naturae* is distilled in the following quote, cited throughout the Middle Ages, and written by fifth-century philosopher Macrobius, one of the last pagan writers of ancient Rome,

> Since, from the Supreme God Mind arises, and from Mind, Soul, and since this in turn creates all subsequent things and fills them all with life . . . and since all things follow in continuous succession, degenerating in sequence to the very bottom of the series, the attentive observer will discover a connection of parts, from the Supreme God down to the last dregs of things, mutually linked together and without a break. And this is Homer's golden chain, which God, he says, bade hang down from Heaven to Earth.[4]

But this Great Chain of Being was, partly, a theory of biology. It was a model of the generation of 'sentient and vegetative creatures'. It was an enormously influential idea, until eclipsed by the theory of evolution, many centuries later. Indeed, and as we shall see, the idea of a Chain 'linked together and without a break' lay at the centre of the nineteenth-century scandal of extinction.

Once again, idealist notions are at the heart of the problem. The influence of Plato and Aristotle reigns supreme in the Great Chain. Plato's principle of plenitude dominates the upper part of the Chain. Plato taught that the cosmos contained all possible forms of existence; everything that could exist, does exist. After all, a cosmos in which some species was not realised would be imperfect, since it would not be full. Hence, the Platonic inspiration is at the heart of the idea that the Most Perfect Being would 'emanate' and 'overflow' into the universe. Forms and copies of its true Self are created, replicating down in descending sequence, to the 'last dregs of things'.

If the upper part of the *scala* was Platonic, the lower part revealed the sway of Aristotle's biology, revived around AD 1200, with its notions of continuity and gradation:

> Nature passes so gradually from the inanimate to the animate that their continuity renders the boundary between them indistinguishable; and there is a middle kind that belongs to both orders. For plants come immediately after inanimate things; and plants differ from one another in the degree in which they appear to participate in life. For the class taken as a whole seems, in comparison with other bodies, to be clearly animate; but compared with animate to be inanimate. And the transition from plants to animals is continuous; for one might question whether some marine forms are animals or plants, since many of them are attached to the rock and perish if they are separated from it.[5]

Always the pragmatist, Aristotle had refused to discuss the ultimate origins of nature. His view was idealist: the world always was as it is, since that is the logical way for it to be. This rather steady state view of the cosmos was to prove problematic when adopted by the Church as its cosmology. So they simply stuck a creation event at the beginning, and a sudden destruction at the end.

Nature's cypher, according to Aristotle, was physics, a very *biological* physics. His notion of nature was a form of animism. It centred on the possibility that souls exist in animals, plants, and people. This was not an account of lifeless matter in motion, but a reading of the world as if everything were actually alive. His account of all things, from why stones fall to why some men are slaves, reduces to the same cause, 'it is their nature to do so'. Aristotle's biology, and its hierarchy, became the standard ancient system of nature, passed on through the centuries.

So Aristotle was fundamental to the Great Chain of Being. The 'will of God' meant creation was a process of degeneration by descent. In this sense, it is the very opposite of the idea of evolution that was to be developed by the likes of Darwin. And yet in another way, the Chain may be thought an evolution of sorts: Aristotle believed that nature strives for perfection. The main thrust, however, was stasis: nothing really changed in nature. Species were eternal signposts on the road to perfection, or imperfection. In fact, imperfection was Aristotle's inclination. His idealism meant he was more likely to see an animal as an imperfect man, and a fish as an imperfect animal, than vice versa.

Beings and beasts in the Islamic Golden Age

A kind of extraterrestrial Great Chain of Being life popped up in *One Thousand and One Nights*, known in English as the *Arabian Nights*. The compendium of Middle Eastern and south Asian folk tales had been compiled in the Islamic Golden Age, between the eighth and thirteenth centuries. The stories were collected over a number of centuries by numerous authors and scholars, across the Middle East, Africa, and Asia. The tales they told had exotic roots, tracing back to folklore from ancient and medieval Arabic, Indian, Turkish, Persian, Egyptian, and Mesopotamian sources, with many tales originating out of folk stories from the Caliphate era.

One particular tale, *The Adventures of Bulukiya*, shows early science fiction elements of the pluralist kind. The story's unsurprising protagonist, Bulukiya, goes on a quest for the herb of immortality. His quest takes him on a voyage that embraces ocean exploration, journeys through Heaven and Hell, and an odyssey across the cosmos to other worlds, radically alien to his own. Along the way Bulukiya encounters various forms of exotic life, including djinns, talking trees and serpents, and mermaids.

The genre of the fantastic voyage harks back at least to the olden Sumerian *Epic of Gilgamesh*. Many ancient tales of fantastic journeys were more or less odysseys of the soul into the afterlife. *The Adventures of Bulukiya* also shows hints of a provenance that stems from the *Gilgamesh* epic. Marvel is heaped upon marvel. True, Bulukiya's quest is for the herb of immortality. But it is a journey on which he encounters speaking serpents as big as camels, capturing the Serpent Queen, walks on the surface of a sea, unearths the tomb of Solomon, visits alien and utopian islands, discovers trees that laugh, trees that bear strange fruit of human heads and birds on their branches, watches mermaids at play, is granted an audience with the King of the Jinn, and is lectured by the King on the nature of the innumerable hells of Allah.

Elements of *Bulukiya* portray a multiplicity of worlds. True, some are diabolical, such as the least of the hells of Allah, Jahannam. In Jahannam, there are one thousand mountains of fire, each mountain housing seventy thousand cities of fire, in each city seventy thousand castles of fire, in each castle seventy thousand houses of fire, in each house seventy thousand couches of fire, and finally on each couch seventy thousand manners of torment. This wonderfully hellish impression of infinity imaginatively anticipates the huge landscapes of twentieth-century science fiction.

Bulukiya receives further instruction on the scale and nature of the cosmos. His next tutor is a huge angel who sits on the peak of the cosmic mountain of Kaf. The angel teaches Bulukiya about the plurality of worlds beyond the one known to him. There are forty such worlds, each more than forty times the size of Bulukiya's world, which is contained within Mount Kaf. Each world has its own strange and alien inhabitants. And each world has its own colour, with its own complement of guardian angels.

The scale of the worlds themselves is dwarfed by the sheer size of the angels who watch over them. And once more the narrative weaves its imagery in polymerised prose: a chain of beings, varying in size. Each angel stands on a rock, which is supported by a bull, which stands on a fish, which swims in the mighty ocean. Bulukiya is permitted to observe this fish. Its head takes a full three days to pass before his eyes, in another reference to the scale of being. And yet the fish is small before the ocean. And the ocean sits above an airy abyss that is above a realm of fire, and beneath all this, beneath this evocation of the elements made gigantic in scale, is the cosmic serpent, which can swallow whole all that is above it, and barely notice the difference.

Clearly, the medieval Muslim conceived himself to be alive in a cosmos that was vertiginous in scale. And the grandeur of the imagined worlds in which Bulukiya journeyed is not unique in Arabic literature in evoking the quintessence of the *aja'ib*, the astounding. There were other Islamic and Gnostic speculations on alien life on other worlds. Witness the eleventh-century philosophical fantasy, *Salaman wa-Absal*. The tale was penned by Ibn Sina, perhaps better known in the West by the Latinised name Avicenna, a Persian polymath who wrote around four hundred and fifty discourses on a range of different topics. His protagonist, Absal, journeys vast deserts and alien seas, meeting bizarre creatures, such as 'humans' with the skins of quadrupeds, and on whom vegetation grows. And yet Absal travels beyond the Earth, and is introduced to alien inhabitants on the Moon and other planets, even travelling to the stars where he encounters the inhabitants of zodiacal cities.

From The Throne of God, to the Meanest Worm

For two thousand years the book of life was slammed shut.

During the Dark Ages of faith, the Great Chain of Being was moulded into a more recognisably Christian form. For Heaven and for Earth, Christian scholars put the Chain to divine rights. One anonymous sixth-century philosopher, the Neo-Platonist known as Pseudo-Dionysius, fixed the order of angels in the upper reaches of the Chain. They kept the heavens in motion it seems. The Primum Mobile was turned by the Seraphim, the fixed stars by the Cherubim, and so on, down to the more lowly angels, who had the misfortune of caring for the Moon.

The great thirteenth-century philosopher St Thomas Aquinas laboured over the Chain's connectivity. The crux, thought Aquinas, was the dual nature of man. Once more, the idealists considered man's place to be central, an anthropocentrism writ large in nature. In the continuity of the Chain, the lowest place of the higher genus is always found to border upon the highest place of the lower genus. This tenet was as true for zoophytes, which are half plant, half animal, as it was for man, who

> has in equal degree the characters of both classes, since he attains to the lowest member of the class above bodies, namely, the human soul, which is at the bottom of the series of intellectual beings – and is said, therefore, to be the horizon and boundary line of things corporeal and incorporeal.[6]

And so medieval scholars extended the Chain, drilling down through the four elements, into the dirt of inanimate matter. Aristotle had been vague about his criteria for degrees of perfection. So where no obvious cyphers told of an object's difference, the black arts of astrology and alchemy showed 'correspondences' and 'influences'. Thus each planet connected with a colour, a metal, a day of the week, a stone, and a plant, each association defining their place in the Chain.

Lowest of all lurked Lucifer. The Chain was projected down into the supposed conic cavities that were the bowels of the Earth. There, legions of devils infested the nine circular slopes, which mirrored the nine heavenly spheres above. Lucifer himself loitered at the gravitational well of this downward cone, the exact centre of the Earth, and the sorry end of the Chain. It was a scale of nature that was not so much geocentric as 'diabolocentric'.[7] A Chain with Hell at its heart, for although continuous and the heavens incorruptible, the Earth sat amongst 'the filth and mire of the world, the worst, lowest, most lifeless part of the universe, the bottom storey of the house'.[8]

In this worldview, the idea of evolution was heresy. The potency and power of the Chain as a vision of animate creation, and man's place in it, is testified in time. For the sacrifice of a break in the chain, no matter where, would be the collapse of the entire cosmic order. Once again, we comprehend the idealist cosmos. It is a static world, impervious to change. A cosmic inventory of creation, cast in permanence by God.

As in astronomy, so it was for biology, in the ancient and medieval world. Man was at the midpoint. Any shift in the cosmic place of Earth and man would imply moving the throne of God itself. Any change in the Chain of Being, however small, would have profound consequences. It might bring the rigid and graded hierarchy of life crashing to the ground. Even before the rest of the tale is told, we can see what the future is going to bring for the Chain.

In biology and society, change was the outlaw. No evolution, and no social progress. A man would seek mobility in this medieval life in vain. One was meant to be consoled by the idea that movement in the Chain was granted only after death. In this corporeal world, ordained rank was final. But inconsistency was rampant within the system. The lowly material world was meant to be one of mutability. Surely change must be possible in such a world. But the Chain brought blessed immutability to Earth, with an inert hierarchy as applicable to human affairs as it was to animal, vegetable, and mineral.

But there are far more important factors than Christian dogma – economy and industry included.

By the mid nineteenth century, great engines of change were turning over the soil of the world. The speculation about the Earth and its fossils steadily grew, along with a great fascination with the natural world. The embryonic sciences of geology and biology began to locate man's place in the depths of time, as astronomy did to chart our position in cosmic space. Together, these sciences spelled out something new too for astrobiology.

Indeed, the word 'biology' entered the lexicon of science. This new science of living things was rather different from the prior practice of natural history, with its emphasis on the three kingdoms of nature: animal, vegetable, and mineral. The *scala naturae* had been key to that natural history tradition. But the new field of biology bore witness that *discontinuity* had replaced the earlier notion of the continuous contours of the Chain, where the most archaic creatures progressively faded into fossils, gems, and minerals.

The new science of geology also revolutionised the world. And like biology, the sway of geology spread far beyond the scientific horizon. Geology helped destroy established truths, forcing the planet to confront the terrible extent of history. And the cosmos in which the Earth sat got more ancient by leaps and bounds. In the face of such terror, biblical literalists sought refuge in the *scala naturae*. But, by design, the Chain was no more robust than its weakest link. Indeed, the very integrity of the Chain was proof of God's glory. There could be no 'missing link', a term later appropriated by the emergent evolutionists. In the words of English philosopher John Locke:

> In all the visible corporeal world we see no chasms or gaps. All quite down from us the descent is by easy steps, and a continued series that in each remove differ very little one from the other. There are fishes that have wings and are not strangers to the airy region, and there are some birds that are inhabitants of the water, whose blood is as cold as fishes . . . When we consider the infinite power and wisdom of the Maker, we have reason to think that it is suitable to the magnificent harmony of the universe, and the great design and infinite goodness of the architect, that the species of creatures should also, by gentle

> degrees, ascend upwards from us towards his infinite perfection, as we see they gradually descend from us downwards.[9]

The account from Locke speaks strongly of the damage done by the idea of extinction in the minds of the pious. 'It is contrary to the common course of providence to suffer any of His creatures to be annihilated',[10] as the Quaker and Royal Society botanist Peter Collinson declared. And, in the words of seventeenth-century English naturalist John Ray, evidence of 'the destruction of any one species', would result in, 'a dismembering of the Universe, and rendering it imperfect'.[11]

The Origin of Species

And yet extinction is exactly what the nineteenth century delivered. The horror of extinction and the associated 'missing links' was to prove fatal. Like comets and sunspots in astronomy, the 'missing links' were monstrous anomalies, fell portents of undesired change. Here was proof that something was very wrong with the Great Chain of Being, a system that was supposedly framed by the hand of God. Astrobiology was about to be transformed. If the discoveries with the telescope were a journey in space, then evolutionary theory was a trip through time. Scientists like Darwin and Wallace, and with them the world of the nineteenth century, were about to experience the ride of a lifetime, a ride in which time itself would seem to warp and telescope.

Suddenly, any Christian interpretation of natural history, including the extraterrestrial, faced problems on the biological front. Just as the telescope had unveiled a cosmos yet unknown, so the fossil record rung like a bell. Its deathly toll resounded with a cornucopia of cadaverous creatures. Fossilised flowers that no one had ever seen bloom, woolly mammoths and rhinos, the terrible lizards, and a whole host of bizarre and wonderful beasts.

Furthermore, the remnants of these extinctions were found in curious locations, places where they could not possibly have prospered: sea creatures on hilltops, polar bears on the equator, that sort of thing. The world must have seen massive upheavals for such profound changes to have occurred in the short time dreamt up by the biblical literalists.

And if dead creatures proved difficult, there were also new *living* ones to contend with. The visceral harvest of the fossil record plagued the same Christian theorists who were then equally mystified by the array of living creatures being discovered. Through the exploration that came with the piracy and plunder in the deep jungles of the globe, naturalists were faced with a burgeoning biota. Countless odious beasts that were so obscure it was hard to see why a divine hand would have brought them into the world at all.

The great discovery of the nineteenth century, in which Darwin played a part, was not merely to assert that life evolved. That had been argued before. Rather, the true innovation of the age was to discover the evolutionary mechanism by which new species came to be. For no small reason did Darwin name his book, *The Origin of Species*. 'Evolution' was not a word Darwin liked to use. 'One may say,' Darwin wrote, 'there is a force like a hundred thousand wedges trying to force every kind of adapted structure into the gaps in the economy of nature, or rather forming gaps by thrusting out weaker ones.'[12] Here was a natural mechanism that was not only global, but also potentially cosmic. It was a means that canvassed continually to snuff out most variations, while keeping those few carried by individuals who had won the struggle to survive and breed. Here was 'natural selection'. The individual differences between members of a species, in tandem with environmental factors, shape the chances that an individual will pass its characteristics on to posterity.

Darwin's *Origin of Species* identified natural selection as the mechanism that creates new species. Twenty-three centuries after Ionian philosopher Heraclitus suggested all was in a state of flux, Victorian science resurrected an Atomist world in a state of constant change. The new science of biology had identified two things. One, that nature favoured variety. And two, that nature prefers geographical spread. For the further afield a species is scattered, the less tied its prospects to a single setting.

Of course, the theory of evolution is not the result of a single worker. From antiquity on, the finest philosophers pondered the rich variety of life. And over those centuries, divine creation had not always been thought of as the causal factor. A thread of eminent thinkers, from Empedocles, through the Atomists Epicurus and Lucretius, to Leonardo da Vinci, had preferred more secular suppositions. Rather than believing in the Great Chain of Being, that every form of life was created by God and has stayed unaltered since, these sages looked to nature's inherent patterning for an answer.

The same was true for philosophers such as Francis Bacon, René Descartes, and Gottfried Wilhelm Leibniz. All favoured natural over supernatural causes for the way in which species changed over time. The French naturalist Étienne Geoffroy Saint-Hilaire had made important progress in the late eighteenth century. He had proposed that certain species adapt and survive when environments change. Others perish, and become extinct. Saint-Hilaire had come the closest to grasping the course of evolutionary change. But he lacked the key explanation of how new species were created, and how the millions of species on planet Earth had emerged from a common ancestor.

So, significant as the publication of *Origin of Species* may be, evolutionary views were already widespread before 1859. The question for us is how the theme of evolution, and natural selection, impacted upon the pluralist debate. How was the mechanism of evolution imagined in a cosmic setting, and what new approaches were inspired to tackle this ancient question of the existence of extraterrestrial life?

To aid in our inquiry for possible answers, we shall, as with other chapters, focus on a selection of key texts that represent and crystallise the thoughts of the age. Here, after Darwin as it were, we shall explore the work of five authors: Charles Darwin, Alfred Russel Wallace, Richard Proctor, Camille Flammarion, and H. G. Wells. Along the way we shall also run into other workers and thinkers in fields and philosophies that impinge upon the pluralist debate.

Before we fall headlong into our cosmic odyssey on evolution, however, it would be well to mention another significant development in science, around the period centred on 1860. The innovation was the spectroscope. Like the telescope before it, the spectroscope utterly transformed the study of the stars. For now, astronomers could establish the chemical make-up of heavenly bodies. The significance came with this chemical signature. Here was the clearest evidence that the elements of which the Earth was made were also present in the cosmos, perhaps the entire universe. And if this were true, then it suggested laws that held good for Earth might also hold true for the heavens.

A 'New Astronomy' was born. Workers in the field began to call themselves astrochemists, and even astrophysicists. In this cultural milieu, the evidentiary aspects of spectrum analysis began to play a major part in the pluralist debate. One of the leaders of this new field was English amateur astronomer, William Huggins. From his private observatory in south London he was the first to take the spectrum of a planetary nebula, and also the first to distinguish nebulae from galaxies, by showing that some spectra were essentially gaseous, while others, such as the great galaxy in Andromeda, were characteristic of stars.

Huggins worked much with British chemist William Allen Miller. Together they published *On the Spectra of Some of the Fixed Stars*, which contains the following passage on the question of extraterrestrial life:

> It is remarkable that the elements most widely diffused through the host of stars are some of those most closely connected with the constitution of the living organisms of our globe, including hydrogen, sodium, magnesium, and iron ... On the whole we believe that the foregoing spectrum observations on the stars contribute something towards an experimental basis on which a conclusion hitherto but a pure speculation, may rest – viz. that at least the brighter stars are, like our sun, upholding and energising centres of systems of worlds adapted to be the abode of living beings.[13]

Darwin, Wallace, and extraterrestrial life

But the major discovery of the age was the theory of evolution. The theory, of course, had at least a co-discoverer in Alfred Russel Wallace. Darwin's junior by a decade, Wallace was a radical and open-minded naturalist. He was self-taught, and both intuitive and unconventional. As a young man Wallace freely roamed the globe. He explored much, from the rainforests of the Amazon Basin, to the Malay Archipelago. It was here he identified the Line that divides Indonesia into two distinct parts: one in which creatures closely related to those of Australia are common, and one in which the species are almost exclusively Asian in origin.

On 1 July 1858, a joint Darwin–Wallace paper on the new theory was read at the Linnean Society of London, the world's premier society for the communication of natural history. Throughout the lifetimes of the two men, both were given credit. But after Darwin's death, a small industry sprang up. Biographies were penned by the patriarch's family, and the Darwin myth created. But some scholars are keen to recognise a more collective effort:

> Although the Linnean Society meeting judged Darwin and Wallace to be equally worthy of recognition as the originators of the theory of evolution, one of them had to be recognised as pre-eminent. The Linnean Society was made up of gentlemen natural philosophers. Wallace, who had written out the complete theory of evolution to which they had listened in silence, was not a gentleman. Charles Darwin, whose unconnected thoughts were contained in two extracts from a 14-year-old essay and a copy of a recent letter, was a gentleman. In the social context of the time, a gentleman always trumps an employee. Thus, the document merging the two presentations referred to the 'Darwin-Wallace' theory of evolution. In the lifetimes of both men (Darwin died in 1881, Wallace in 1913) it was usual for the theory to be referred to by this title, but after Wallace's death it became 'Darwin's theory of evolution'. For almost a century, Wallace's scientific achievement has been effectively buried under Darwin's reputation.[14]

Darwin and Wallace also had contrasting attitudes to the question of extraterrestrial life. Together their work gifted a potent argument for the pluralistic debate, providing a deeper materialist understanding of the cosmos, and which supported many advocates of pluralism. Teleology was challenged. Whither design when natural selection provided a natural mechanism for the creation and development of life off, as well as on, Earth? Darwin's work is characteristically and dully bereft of such mentions of the extraterrestrial dynamic. He was even conservative in referring to his published contribution to evolution as the *Origin of Species*, rather than the *Origin of Life*. But Darwin did make brief mention of his views on the origin of life, albeit in a terrestrial setting. The quote comes from a letter he wrote on 1 February 1871 to Joseph Hooker, one of the leading British botanists and explorers of the nineteenth century:

> It is often said that all the conditions for the first production of a living organism are now present, which could ever have been present. But if (and oh! what a big if!) we could conceive in some warm little pond, with all sorts of ammonia and phosphoric salts, lights, heat, electricity, &c. present, that a proteine compound was chemically formed ready to undergo still more complex changes, at the present day such matter would be instantly devoured or absorbed, which would not have been the case before living creatures were formed.[15]

Darwin's 'warm little pond' seems comically quaint in light of today's talk of matter in the universe being born in violence. Nonetheless, it remains a competing scenario for life's origins, and still has a significant bearing on the pluralist debate.

Before we consider Wallace's views on extraterrestrial life, consider the wider context.

From its germination in the minds of naturalists such as Darwin and Wallace, the theory of evolution was destined to develop into an ideological as well as a scientific conflict. Philosophically, the theory cast Plato and Aristotle into the dustbin of history; the Great Chain of Being was a broken chain, its elusive links dissolving into the ether. Evolution also finally dismantled any validation of Aristotle's idea of purposeful final causes. No wonder that many theologians, whose worldview was finalistic, condemned evolution with hysterical fervour.

The other context worth considering is in the form of the two-culture split, famously expounded a century later by British scholar C. P. Snow. In the 1959 Rede Lecture, Snow introduced into the general lexicon a shorthand for the difference between two attitudes, that of the sciences and that of the humanities. Snow characterised the attitude of *science* as one in which the observer can objectively make unbiased and non-culturally embedded observations about nature. In contrast, he characterised the attitude of the *humanities* as a worldview in which science was seen as embedded within language and culture. Importantly, Snow's lecture was later published under the title *The Two Cultures and the Scientific Revolution*.

But the two-culture split was already there in the days of Darwin and Wallace. Humanists like Wallace, along with the literary and artistic movement, discarded those aspects of science they felt had coupled up with the machine age, and everything in its train. The outcome was to prevent any cooperation between the two branches of intellectuals. One can only imagine the wary scepticism with which Darwin would indulge the views of Wallace. Mainstream scientists were blunted by a quite deliberate rejection of anything, and especially art and poetry, that did not come into the ken of their, by now, rather specialised work. It also explains Darwin's reluctance to speculate on the question of extraterrestrial life.

We are dealing here with two very different vectors of science in Darwinism and, what one might call, Wallaceism. Rather than Darwin's radical materialism, Wallaceism was a different kind of evolutionary theory, one immersed in humanism. By 1864, Wallace had published a paper, *The Origin of Human Races and the Antiquity of Man Deduced from the Theory of 'Natural Selection'*. It was his application of evolutionary theory to man himself. Darwin had not yet communicated the subject in public, though Thomas Huxley, Darwin's 'Bulldog', had done so in *Evidence as to Man's Place in Nature.*

Not long after, Wallace turned to spiritualism. He held that natural selection did not explain scientific, or artistic, genius. Nor could it explain metaphysical musings, wit, or humour. Wallace believed 'the unseen universe of Spirit' had mediated at least three times in history. The first occasion was the generation of life from the primeval soup of the early Earth. The second time was the development of consciousness in higher animals. And third was the maturity of the higher mental faculties in man.

Wallace held that the *raison d'être* of the cosmos was the evolution of the human spirit. He would not have it that human intellect could be explained away by natural selection. A higher spiritual force simply had to be at work. When Darwin read Wallace's thoughts on the matter in the April 1869 issue of *The Quarterly Review*, he viewed them with disdain, and scratched a tetchy 'NO!!' in the margin of his copy.

This contrast between the rational Darwin and the spiritual Wallace has an antecedent in our tale. Consider Galileo and Kepler. Galileo seems utterly and worryingly modern. But Kepler seems never to have severed himself from the mystical Middle Ages. Galileo seems devoid of any spiritual leanings. Kepler seems struck by the magical implications of science. Like Wallace, Kepler's science endeavoured to lay bare the secrets of the universe. They were a hodge-podge of geometry, music, astrology, astronomy, and the occult. And yet, Kepler conjured up the three laws of planetary motion, a strong belief in pluralism, and the first modern work of extraterrestrial fiction. From a similar standpoint, Wallace divined the theory of evolution through natural selection.

The paths of Wallace and Darwin soon diverged. Darwin's strident materialism found its way into two books, *The Descent of Man* (1871) and *The Expression of the*

Emotions in Man and Animals (1872). Darwin's subject was the evolution of human psychology and its connection with the animal behaviour. His conclusion was

> that man with all his noble qualities, with sympathy which feels for the most debased, with benevolence which extends not only to other men but to the humblest living creature, with his god-like intellect which has penetrated into the movements and constitution of the solar system – with all these exalted powers – man still bears in his bodily frame the indelible stamp of his lowly origin.[16]

Both books were well received, and very popular.

But Wallace's evolutions remained in spiritual vein. He created an alternative kind of evolutionary theory, a type of teleological evolution. Wallace's was an evolution of purpose and design, encapsulated in a theory that held all things were designed for, or directed toward, a final outcome. As Wallace's biographer Martin Fichman maintains, 'Wallace's emerging evolutionary worldview was compatible with a broader spiritual and teleological framework that would become more overt on his return to England.'[17] According to Fichman, Wallace's observations of orang-utans led him 'to invoke more explicitly the concept of design in nature'.[18]

Wallace, of course, was struck down by idealism. The respective evolutions of Darwin and Wallace after the publication of *Origin* help expose that their choice was embedded in that ancient historical struggle, part-scientific, part-political. Namely, the centuries-long duality of philosophy between materialism and idealism. For the Darwinians, Wallace's peculiar strain of evolutionism would only have endangered the materialist theory, and ruined years of promotion for the superiority of natural selection in the world of science.

Not all evolutionists believe in extraterrestrial life. Even though a scientist like Wallace may have understood evolution by natural selection, that did not necessarily make him an advocate of pluralism. In the twentieth century, perhaps the world's most renowned evolutionary biologist, Ernst Mayr, also challenged the existence of extraterrestrial life and intelligence, though on somewhat different terms. During the late machine age, Wallace, already in his seventies, began to survey the sciences. His analysis was published as a book entitled *The Wonderful Century* (1898).

The review of astronomy that Wallace carried out led to the publication of a controversial article under the heading of 'Man's Place in the Universe'.[19] It was later extended into a much fuller account, the book *Man's Place in the Universe: A Study of the Results of Scientific Research in Relation to the Unity or Plurality of Worlds*. In both publications, Wallace clashes with the contemporary consensus by suggesting that recent scientific progress, particularly astronomy, made it more than likely that man's place in the cosmos 'is special and probably unique'.[20] Not only that, but 'the supreme end and purpose of this vast universe was the production and development of the living soul in the perishable body of man'.[21]

A bold claim indeed. The concluding passage of Wallace's work was no less courageous:

> The three startling facts – that we are in the centre of a cluster of suns, and that that cluster is situated not only precisely in the plane of the Galaxy, but also centrally in that plane, can hardly now be looked upon as chance coincidences without any significance in relation to the culminating fact that the planet so situated has developed humanity.[22]

Contemporary astronomy did indeed seem to support Wallace's first couple of contentions, but 'Man's Place', both essay and book, caused a major stir. In a bid to strengthen his argument in the face of his critics, Wallace added an appendix to his book, which set out a case based on 'organic evolution'. Here he presents a critique of those who, on the grounds of evolution, 'consider it absurd ... to suppose that man, or some being equally well organised and intelligent, has not been developed many times over in many of the worlds they assume must exist'.[23] In answer to this argument, Wallace, co-founder of evolutionary theory, claimed that such scholars do not

> give any indication of having carefully weighed the evidence as to the number of very complex and antecedently improbable conditions which are absolutely essential for the development of higher forms of life from the elements that exist upon the earth or are known to exist in the universe. Neither does any one of them take account of the enormous rate at which improbability increases with each additional condition which is itself improbable.[24]

Stating his case further, Wallace draws attention not only to the number of conditions necessary for the evolution of intelligence and the improbability of each condition, but also that they must concurrently be present and act over huge swathes of time. Indeed, so convinced was Wallace of his case that he stated, 'the total chances against the evolution of man, or an equivalent moral and intellectual being ... will be represented by a hundred million of millions to one'.[25]

One wonders at the origin of Wallace's antipluralism. In his work he refers to the Christian case against pluralism, but also contends,

> Whether we look at this great problem from the agnostic or religious point of view; whether we study it as scientific monists or as philosophical spiritualists, our only safe guide is to be found in the facts

> and the laws of nature as we know them, and in the conclusions to be logically derived from an unbiased application of those laws to the question at issue.[26]

With Wallace, we have the first evolutionist sceptic.

True, his ideas are strongly mixed with idealism, but much of his writing recognises that contemporary astronomers were guilty of a deterministic view of the possibility of extraterrestrial life. Consider Wallace's words:

> Those among my critics who have expressed adverse opinions . . . give no reasons for this view other than the enormous number of suns that appear to be as favourably situated as our own, and the probability that many of them have planets as suitable as our earth for the development of human life.[27]

Wallace identifies physical determinism very early in the history of science. This determinism centres on the purely physical forces in the cosmos. The idea that the sheer number of stars and orbiting planets is statistically sufficient to suggest other Earths lie waiting in the vastness of deep space.

Other worlds than ours

Influential US astronomer and communicator Carl Sagan recognised that Wallace introduced a much-needed biological perspective to the debate. Sagan admitted to being 'astounded' by the ingenuity and foresight shown in Wallace's work on pluralism.[28] Carl Sagan achieved world fame in the late twentieth century for his science communication, especially in astronomy. One of his near equivalents in the nineteenth century was English astronomer Richard Proctor. When Proctor died prematurely at the age of 51 in 1888, *The Times* recognised him for having 'probably done more than any other man during the present century to promote an interest among the ordinary reading public in scientific subjects'.[29] On the other side of the Atlantic, US astronomer C. A. Young was of the same opinion on the question of Proctor, 'As an expounder and populariser of science he stands, I think, unrivalled in English literature.'[30]

In the twentieth century, Carl Sagan was the author, co-author, or editor of over twenty books. In the late nineteenth century, Proctor had authored fifty-seven books. He was a phenomenon in the popularisation of astronomy in the English-speaking world. His most productive period came in the twenty-five years between his first book and his untimely death in 1888. Most of his published work was in the form of essays that were previously published in periodicals, though Proctor assured his readers they represented less than a quarter of his total output.[31]

Proctor was not known for merely sitting in his study. He went on many speaking tours. He lectured in the United Kingdom, Canada, Australia, New Zealand, and the United States. When he wed an American widow, Proctor moved to America, where he edited the London science journal *Knowledge*, which he had established in 1881. Most significantly, Proctor was not merely prolific. He was the most popular and widely read writer on pluralism in the United Kingdom and United States between 1870 and 1890.

As a result of a financial crash, Proctor, a graduate of St John's College, Cambridge, was pressed into considering a career in science communication. The first of his books, *Saturn and Its System* (1865), was well received by astronomers, but did not bring in the kind of money Proctor was hoping for. With an eye on financial prizes, he wagered that the question of extraterrestrial life might do better. His *Other Worlds Than Ours* (1870) did the trick. From his writing in 1878, it is apparent that Proctor understood the reading public: 'The interest with which astronomy is studied by many who care little or nothing for other sciences is due chiefly to the thoughts which the celestial bodies suggest respecting life in other worlds than ours.'[32]

Proctor's plan proved fruitful. *Other Worlds Than Ours* was in print for over three decades and was in all probability his bestseller. Indeed, Proctor caught the pluralist bug. Having found his niche, and the public's thirst for matters extraterrestrial, Proctor injected a flavour of pluralism into a dozen later books. The change in emphasis was the magic formula. In 1873, one writer was to declare, 'Ten years ago, the name of Richard Anthony Proctor was absolutely unknown; five years later, it was familiar in scientific circles in London; and to-day it is familiar as household words to every educated man in England, and to many thousands in this country [US].'[33]

Proctor proved a worthy and sceptical pluralist. Eschewing the antipluralist stance of William Whewell, his position on Jupiter was indicative, 'Surely no astronomer worthy the name can regard this grand orb as the cinder-centred globe of watery matter so contemptuously dealt with by one who, be it remembered thankfully, was not an astronomer.'[34] Proctor's opinion regarding the systems of Jupiter and Saturn broke new ground. Rather than accepting the traditional pluralist notion of gas giant habitability, he rejected the idea that Jupiter is 'at present a fit abode for living creatures'. Rather, Proctor held that Jupiter is, 'in a sense a sun ... a source of heat', which serves its satellites on which 'life – even such forms of life as we are familiar with – may still exist'.[35] Indeed, as well as presciently suggesting Jupiter was a failed sun, Proctor pressed that the gas giant 'must be intended to be one day the abode of noble races'.[36] In a later publication, he was to reveal the provenance of his new theory on the Jupiter system; he

> set out with the idea of maintaining ... that all the eight known planets ... are inhabited worlds. But even as I wrote that work I found my views changing. So soon as I began to reason out the conditions of life in Jupiter and Saturn, so soon I began to apply the new knowledge which would, I thought, establish the theory that life may exist in these worlds, I found the ground crumbling beneath my feet. The new evidence ... was found to oppose fatally ... the theory I had hoped to establish.[37]

Such ideas portray Proctor as an evolutionist. Perhaps more than most machine age pluralists, he regarded the planets as evolving entities, bodies that were subject to mutability over the course of cosmic time. Moreover, Proctor's ideas themselves were evolving. He often changed his mind, in the face of evidence, on questions of extraterrestrial life, in a series of essays between 1870 and 1875. Thus, Proctor was habitually ahead of the game. Already in 1870, in his *Other Worlds*, his scepticism stood out against mainstream pluralism, which argued habitability of all the planets. And though the book was reprinted twenty-nine times between 1870 and 1909, Proctor had changed his mind, or even abandoned, many major points by 1875.

Mars is a case in point. By the end of the century, and influenced by Proctor himself, life in the Solar System was regarded as more exceptional. The consensus of planetary habitability had been whittled down to Earth and Mars, with the red planet being identified as the only immediate hope for finding extraterrestrial life beyond Earth. Mars, of course, also became the focus of considerable controversy at the end of the machine age, with the debacle associated with the assumed presence of canals on its surface.

Of Mars, Proctor had written in *Other Worlds Than Ours* that it 'exhibits in the clearest manner the traces of adaptation to the wants of living beings such as we are acquainted with. Processes are at work out yonder in space which appear utterly useless ... unless ... they subserve the wants of organised beings.'[38] Though Proctor's early work exhibits a tendency towards teleology, he nonetheless supports his Mars contention with a little evidence. He cites reports of water vapour in the Martian atmosphere, along with observations of oceans, ice caps, and seas.

And yet by 1873 his view had changed dramatically. Proctor's book *The Borderland of Science* contained a paper entitled 'A Whewellite Essay on the Planet Mars'. Herein, Proctor asks what a contemporary Whewellite might argue against life on Mars. Proctor sets out such a case. The red planet is set further out from the central Sun, so it receives less heat than Earth. Because of its smaller mass, it has less atmosphere and keeps less heat than Earth. And so, he concludes, 'Neither animal nor vegetable forms of life known to us could exist on Mars.'[39] Not only that, but if living beings do exist on that world, they must, 'differ so remarkably from what is known on earth, that to reasoning beings on Mars the idea of life on our earth must appear wild and fanciful . . .'[40]

This Whewellite approach was developed further in two books Proctor had published by 1876, *Our Place Among Infinities* and *Science Byways*. In the latter of these two books, Proctor outlines his evolving and sceptical strategy on the question of extraterrestrial life, proclaiming at one point, 'Millions of uninhabited orbs for each orb which sustains life!'.[41] Nonetheless, he still felt free enough to declare it 'at least as probable that every member of every order – planet, sun, galaxy, and so onward to higher and higher orders endlessly – has been, is now, or will hereafter be, life-supporting "after its kind".'[42]

The source of such flexible confidence in his own opinion on pluralism was Proctor's new theory of planetary evolution:

> Each planet, according to its dimension, has a certain length of planetary life, the youth and age of which include the following eras: – a sunlike state; a state like that of Jupiter or Saturn, when much heat but little light is evolved; a condition like that of our earth; and lastly, the stage through which our moon is passing, which may be regarded as planetary decrepitude.[43]

In keeping with the developing narrative of the machine age, Proctor's planetary approach centred on new ideas of time. Christian scholars had previously guessed at the age of the world by adding up the begats. In this fashion, Newton had estimated, given the begats between Adam and Abraham, that the date of Creation was a mere 3998 BC. Kepler had guessed at 3993 BC. This unempirical approach was famously taken to its most farcical apotheosis by James Ussher, the Archbishop of Armagh, who had estimated, 'the beginning of time . . . fell on the beginning of the night which preceded the 23rd of October, in the year . . . 4004 BC'.[3]

Proctor's focus on time was the crucial factor in his new position. He was well aware that the new sciences of geology and biology, and their new concept of time, were revolutionising the machine age. So much so that in 1851, John Ruskin was forced to declare, 'If only the Geologists would let me alone, I could do very well, but those dreadful Hammers! I hear the clink of them at the end of every cadence of the Bible verses.'[3] So Proctor used the new notion of the vast age of the universe to consider the planets as evolving entities over huge swathes of time.

The silent temporal hand of evolution shaped Proctor's later work on pluralism. Evolutionary theories of the Solar System, both nebular and not, influenced his view of the planets. And Darwinian dynamics persuaded Proctor to realise that the eventual depletion of terrestrial coal and energy reserves would mean the entropic decay of the Earth into a desert wasteland like the Moon. He also realised that pluralists were forever applying terrestrial analogy to other worlds, without sufficiently appreciating the age of the Earth itself.

Proctor was also critical of arguments based on design. He rejected the teleological arguments of Chalmers, Dick, and Whewell, reasoning that God's design for the cosmos cannot be known with any degree of certainty. He cites the fact that as 'less than 230 millionth'[44] part of the Sun's energy is incident on the planets, the universe must be replete with such evidence of waste. Hardly good design. By now Proctor had banished Solar System life from any moon or planet other than Earth. On Mars it had disappeared, though, 'the development of higher forms of life may have been less complete than on our earth'.[45] Jupiter lay in wait. Lifeless still, it was evolving an enabling habitat, one that would embrace 'creatures far higher in the scale of being than any that have inhabited, or may inhabit, the earth'.[46] For Proctor, in time even the Sun will harbour life.[47]

The dialectic of Proctor's position is fascinating. He recognises, in Whewell's idealist and teleological approach, a sufficiently sceptical method that is fit for a materialist to proceed with caution, once stripped of its idealist and disingenuous pretension:

> Have we then been led to the Whewellite theory that our earth is the sole abode of life? Far from it. For not only have we adopted a method of reasoning which teaches us to regard every planet in existence, every moon, every sun, every orb in fact in space, as having its period as the abode of life, but the very argument from probability which leads us to regard any given sun as not the centre of a scheme in which at this moment there is life, forces upon us the conclusion that among the millions on millions, nay, millions of millions of suns which people space, millions have orbs circling round them which are at this present time the abode of living creatures.[48]

The sceptical arguments are weighed on their material merit, and yet the conclusion is this: the infinities of time and space outweigh most parochial conditions. The extraterrestrial life hypothesis wins out. Though this position of Proctor's could again be considered to be physically determinist, a more evolutionary pluralist position endured through the remainder of his published works.

The 1870s also saw Proctor flirting with Catholicism. In 1875, however, he broke off the relationship. The tension between religious idealism and material ideas proved too strong, '[Proctor] severed his connection with that faith on the ground … that church theologians had told him that some of his theories and scientific views were not in conformity with loyalty to the church. He was so convinced of the truth of his ideas, that he left the church.'[49]

His 1877 book, *Myths and Marvels of Astronomy*, saw Proctor sharpen further his pluralist credentials. He outlines clearly how the dialectic of material evidence, such as spectroscopic methods and more refined telescopes, prohibit the more speculative indulgences of yesteryear, but enable other more grounded and evolutionary possibilities:

> If men no longer imagine inhabitants of one planet because it is too hot, or of another because it is too cold, of one body because it is too deeply immersed in vaporous masses, or of another because it has neither atmosphere or water, we have only to speculate about the unseen worlds which circle round those other suns, the stars; or … we can look backward to the time when planets now cold and dead were warm with life, or forward to the distant future when planets now glowing with fiery heat shall have cooled down to a habitable condition.[50]

The most pluralist publication from the last decade of Proctor's life was his *Poetry of Astronomy*, published in 1881. The book included an essay titled 'Is the Moon Dead?' Not only does Proctor explore whether the Moon ever had life, but he also distinguishes between habitability and habitation, a difference rarely drawn out at the time. On a more philosophical level, the volume also finds Proctor in cosmic form:

> May there not be a higher order of universe than ours, to which ours bears some such relation as the ether of space bears to the matter of our universe? And may there not, above that higher order, be higher and higher orders of universe, absolutely without limit? And, in like manner, may not the ether … be the material substance of a universe next below ours, while below that are lower and lower orders of universe absolutely without limit?[51]

Proctor's influence was huge. Countless readers were introduced to the ideas of pluralism through his publications. More importantly, perhaps, was the fact that he popularised an evolutionary and dialectic approach to the topic, partly as a reaction to the idealist and antipluralist position of Whewell, who was nevertheless also a sceptic. For Proctor, planets, moons, stars, and even universes were seen as evolving entities. His positive view of extraterrestrial life was the shape of things to come. He remained a pluralist, even when it became clear that the other planets of the Solar System were probably lifeless.

A brief excursion into subterranean alien life

As the century wore on, the remit of science extended into all domains. Regions of space previously considered beyond the realm of man were now open to exploration, particularly in the imagination. And one of those domains was the Earth itself. In Aristotle's cosmos, made flesh by the Church, Hell itself was at the centre of the Earth. It was locus of the Devil and his legions. It housed the lowliest and most corrupt place in the entire universe.

Aristotle had pictured an unchanging cosmos of nested crystalline spheres each made of ether, the fifth element. As we have seen, the lowly Earth was placed at its centre, the only region of space subject to change and decay. The Church had gone a step further. Dante's *Divine Comedy* had described a journey through the Christian universe. Starting from the planet's surface, the quest had descended into the Earth's bowels, through circles of Hell populated by the sinful dead, and ended at the Earth's core. Dante's subterranean vision mirrored Aristotle's universe above.

The notion had stuck, for some at least, until the eighteenth century. One theologian had suggested that the Earth's rotation was a result of the damned scrambling to escape Hell. The idea of nested spheres persisted in scientific circles. Newton's friend, astronomer Edmond Halley, proposed in a paper published by the Royal Society in 1692 that numerous rotating globes inside the Earth caused its magnetic field.

The first striking use of Halley's idea came in the form of Ludvig Holberg's *Nicolaii Klimii iter Subterraneum* (1745). Translated as *A Journey to the World Underground*, or *Niels Klim's Underground Travels*, the tale tells of a young Norwegian who stumbles down into the Earth to discover an inner planet populated by intelligent nonhuman lifeforms.

The novel starts when Niels Klim returns from the university in Copenhagen, where he studied philosophy and theology. His scientific curiosity drives him to investigate a strange cave hole in the Earth, on the mountain above the town. Niels ends up falling down the hole, and after a short while finds himself floating in free space. The Earth, it seems, is hollow, and populated with an array of alien, but intelligent creatures.

After a few days of orbiting a planet within the hollow Earth, Niels finds that the planet revolves around an inner Sun; it is a Copernican system of inner worlds. Niels is set upon by a gryphon, and falls down onto the planet, Nazar. On the planet, Niels is eventually taken prisoner by tree-like creatures, accused of attempted rape on the town clerk's wife, and put on trial. The case is dismissed and Niels is set the task of learning the language of this utopian state, Potu.

Figure 12 Niels Klim's descent to the Planet Nazar.

During the book, Klim chronicles the colourful culture of this Potuan utopia. He details their religion, their many countries, and their way of life. For example, the Potuans believe that if you perceive a problem at a slower rate, the better you understand and solve it. He is appalled at their gender equality; men and women share the same kind of jobs. So he tables a request to the Lord of Potu to remove women from higher positions in society. For his ignorance, he is exiled to the inner rim of the Earth's crust where he finds a country inhabited by sentient monkeys. Later still, Niels becomes emperor of a land inhabited by the only creatures in the Underworld that look like humans. Eventually he falls up through the crust and back up to Bergen, Norway, the place of Holberg's birth. Like Cyrano de Bergerac before him, Holberg imagined meetings between man and alien. His is a universe fit for life, even if it does reside within a hollow Earth. And the novel can be read as a satirical attack on human self-esteem, and the assumed pre-eminence of humanity.

As machine-age geology developed and the death roll of extinction grew, the terrible extent of history began to dawn. Jules Verne's *Journey to the Centre of the Earth* (1864) wielded the new geology like a club. Verne's creative journey had begun in 1863 with the first of 63 *Voyages Extraordinaires: Voyages in Known and Unknown Worlds*. An early advert claimed Verne's goal was 'to outline all the geographical, geological, physical, and astronomical knowledge amassed by modern science and to recount, in an entertaining and picturesque format that is his own, the history of the universe.'[52] Some mission.

Journey to the Centre of the Earth is classic Verne. The machine age had its first great sceptic of science in Mary Shelley and *Frankenstein*, but its chief positivist in Verne. Though the mythology of the machine is absent, the penetrative thrust of science is unmistakable. Verne's book promotes a heady confidence in progress, and portrays a predictable cosmos in which the unknown is easily assimilated into our taxonomies.

The feeling of estrangement in science fiction is bound to the scientific worldview, and the alienating discovery of life in the new universe. Though this separation from nature began with Copernicanism, it reached its peak in the machine age. The sheer pace of dizzying change was a key factor. So too was the Victorian crisis of faith hastened by the emergent sciences of biology and geology.[53] The modern age of alienation had truly begun. The science fiction of the age can be seen as an attempt to repair this sudden separation from nature, to reload the emptiness, to somehow jack-in to the void.

Verne's book is a voyage through a subterranean world, and a conquest of space. The novel's paradigm is this: nature is a cypher to be cracked. Gone are Dante's mythical speculations of an earthly core, locus of the Devil and his legions. In its place is a quest into the depths of evolutionary time. The trail blazed by geologists such as Hutton and Leclerc had also inspired the palaeontologists. George Cuvier had anticipated the idea of species extinction. Once in the subterranean caverns, grottos, and waters, Verne's explorers find the interior alive with prehistoric plant and animal life. A herd of mastodons, giant insects, and a deadly battle between an Ichthyosaurus and a Plesiosaurus follow.

A giant prehistoric man found overlooking the mastodon herd is another of Verne's nods to contemporary science. When Verne's professor lectures on the latest anthropological discoveries, he refers to Boucher de Perthes, who in 1863 had unearthed a human jaw in northern France, suggesting man was over a hundred thousand years old. Verne waited until the discovery was confirmed before including it in his 1864 novel. Significantly, this panorama is subjected to an orgy of classification at the hands of Verne's travellers. To name is to appropriate and conquer. Their taxonomy is an attempt to bleed nature of its strangeness, to render it human.[54]

When Darwinism provided writers with the metaphor of evolution, around seventy futuristic fantasies were spawned in England alone between 1870 and 1900. One of the first, Edward Bulwer-Lytton's *The Coming Race* (1871), was a book about subterranean supermen. His fascinating, if bizarre, tale is set in an underground world of well-lit caverns. As with Ludvig Holberg's story, Lytton's book begins as the narrator falls into an underground hollow. Their heroes, it seems, are unduly careless.

Nonetheless, a mysterious subterranean human-like race is unearthed. They derive immense power from *vril*, an all-permeating fluid that has enabled them to master nature. *The Coming Race* marked the start of the Victorian obsession with evolving society. In Bulwer-Lytton's book, a new kind of life has been secured through the application of science. The novel suffers from machine determinism, however. The new technology has no social agency, even though it is described as having inevitable social consequences. Far-flung subterraneans who do not have *vril* are uncivilised (or, we might say, 'the great unwashed', since the aristocrat Lytton is alleged to have coined the phrase). Indeed, the possession of *vril* energy is the civilisation, and the refinement of society based on technology alone.

The book strikes one final fearful note. As suggested by the ominous title, once the more advanced Vril-ya surface from their caverns, they will take the place of man: 'the more deeply I pray that ages may yet elapse before there emerge into sunlight our inevitable destroyers'.[55] Meanwhile, *The Coming Race* proved to be truly inspiring for a Scottish industrialist; he made a fortune from a strength-giving beef extract elixir known ever since as Bovril.

La Pluralité Des Mondes Habités

If book publications count for anything, then consider this. If one ponders the history of astronomy publications up to the year 1881, then, of all authors who had ever lived, Richard Proctor ranked seventh in number of publications. He was beaten in fifth place by a French writer who was described as doing 'more toward popularising the study of astronomical science than any of his contemporaries'.[56] It gets better. When one considers the *rate* of publication up to 1881, in that ranking, Proctor comes out third, and his French contemporary first.[57]

The rise of French astronomer Camille Flammarion was meteoric. His story begins with Flammarion, still only a twenty-year-old student in his fourth year at the Paris Observatory, publishing his own ground-breaking account of extraterrestrials. The year was 1862. The book was *La Pluralité Des Mondes Habités*. *La Pluralité* was a short book, barely over fifty pages. But it was an instant hit.

As Flammarion was later to declare, the book 'at once made my reputation'.[58] By 1864, he had extended his book on the plurality of worlds to almost six hundred pages. By the following year the book had garnered twenty-four reviews. And over the next twenty years, thirty-three editions of *La Pluralité* were published. It was a clear indication of the popularity of both Darwinism and the idea of extraterrestrial life. Flammarion argued with enthusiasm that alien life, originating spontaneously rather than divinely, evolved through natural selection in an extraterrestrial setting.

In a sense, Flammarion took over from Proctor. As Canadian astronomer Simon Newcomb would later describe it, Flammarion at first 'wrote so much like a French Proctor that, could a man have a legal copyright on his own personality, the Englishman might have brought suit on the ground of infringement'.[59] Indeed, Proctor presented evidence in his *Poetry of Astronomy* that his French equivalent was plagiarising his writings.[60]

The decade of the 1880s is significant. It saw the death of Proctor, but was only the second of six decades of Flammarion's career. He was the author of over seventy books, with *La Pluralité* continuing in print until the 1920s. Through numerous translations his influence spread beyond his home country. Not until his death in 1925 did Flammarion give up his position as one of the world leaders in the extraterrestrial life debate.

Flammarion's enthusiasm for extraterrestrial life was clear from the very start. Born in Montigny-le-Roi, Haute-Marne, France, at the age of ten he began his four years of study at a seminary. Flammarion was already attracted to astronomy when he spied 'mountains in the moon, as on earth! And seas! And countries! Perchance also inhabitants!'.[61] By the age of fourteen the Flammarion family had moved to Paris. Though he worked for a short time as an engraver, he was 'above all, taken up with cosmological questions, and wrote a big book on the origin of the world'.[62] The book was a huge five hundred pages, and entitled *Cosmogonie Universelle*. It was brought to the attention of Urbain Jean Joseph Le Verrier, best known for his part in the discovery of Neptune in 1846. Director of the Paris Observatory, Le Verrier arranged for the sixteen-year-old author of the book to work under him as an apprentice astronomer.

Then came *La Pluralité*. Flammarion's influences in writing the book bear witness to his flair for popular astronomy. Finding the narrow astrophysics of Le Verrier rather tedious and confining, his passion turned to planets, stars, and pluralism. He was consumed by the writing of past pluralists, including Fontenelle, Cyrano de Bergerac, Huygens, Voltaire, and Herschel. His time at the Observatory rang another change in Flammarion. He lost his Catholic faith due to its irreconcilability with Copernicanism.

Flammarion's imagination was captured by the work of French socialist philosopher Jean Reynaud. Reynaud had been trained at the École Polytechnique and taught at the Écoles des Mines. He had contributed to a number of encyclopedias and was French undersecretary of state in 1838. But in 1854, Reynaud set out his religious philosophy, which he believed to be compatible with Christianity. In his *Terre et Ciel* (1854) he proposed a variation on the theme of the transmigration of souls. After death, the spirit soars and passes from planet to planet, gradually improving with each journey. It was a doctrine of 'indefinite perfectibility',[63] and implied a lowly status for life on Earth. But the notion appealed to Flammarion. In his later remarks on the mystical pluralists of the machine age, he commented, 'Of all the works written on this subject during the period, the most important is without doubt that of our master and friend, Jean Reynaud'.[64]

And so Flammarion embarked upon *La Pluralité*, 'I consecrated the year 1861 to this composition, enflamed with a fiery ardour as one has at age nineteen, not doubting for an instant that I would demonstrate to myself that my conviction in extraterrestrial life was well founded'.[65] The philosophy Flammarion adopted for his first book emerged as 'for me the apotheosis of astronomy and its supreme end'.[66] Indeed, it mushroomed into 'the program of all my literary and scientific life'.[67]

La Pluralité sets out Flammarion's stall. The book is arranged into sections headed historical, astronomical, and physiological, and promises a truly scientific approach. The aim of the history section was to show that 'the heroes of thought and of philosophy have ranged themselves under the banner which we are going to defend'.[68] By juxtaposing the transmigration of souls alongside the existence of alien life, Flammarion suggests the pluralist tradition begins 'contemporaneously to the establishment of man on earth',[69] and is championed by the Egyptians, Greeks, Indians, and Arabs. Flammarion had read widely even at such a young age. Recognising the Greek origin of much pluralist thought, he grants notable status to the Ionian and Atomist philosophers. He is even forgiving of Aristotle. His exception to the allegedly pervasive Greek belief in aliens, Flammarion excuses that, 'had [he] known the true system of the world',[70] the ancient philosopher would have agreed with the rest.

The history section of *La Pluralité* proceeds through the likes of Bruno, Galileo, Newton, and Kant, with Flammarion name-dropping forty-seven prominent pluralists from the seventeenth and eighteenth centuries in just one paragraph. The weight of such opinion, along with quotes from authors such as Herschel and Laplace, helps Flammarion argue that antipluralism is futile in the face of such opposition:

> the eminent men of all ages who have been versed in the operations of nature have also been profoundly impressed by its prodigious fecundity and have understood the insanity of those who would limit that fecundity to our abode alone.[71]

And so to astronomy. Here Flammarion suggests, 'the earth has no marked pre-eminence in the solar system of such a sort for it to be the only inhabited

Figure 13 The enigmatic 'Flammarion Woodcut' originates with Flammarion's 1888 *L'Atmosphère: Météorologie Populaire.*

world'.[72] Rather under-represented in moons, petite in mass and magnitude, Earth does not fare well in comparison with gas giants Jupiter and Saturn, and especially the Sun. Rejecting the idea that the giant planets are poorly placed, Flammarion even so admits the atmospheres of the other planets are 'essentially different'[73] from those on Earth as there is no evidence extant to show they are 'of a chemical composition analogous'[74] to our planet.

The sceptical approach to pluralism is also in evidence in Flammarion's section on physiology. Here he implores his readers not to presume that terrestrial physiology governs life on other worlds. The perpetrators of such a line are charged with 'hurling a gross insult in the shining face of the infinite Power who fashions the worlds'.[75] He does accept, however, that the 'inexhaustible fecundity of nature'[76] evidenced by life on Earth will also be applied on the planets, where Flammarion imagines nature acting with the same wonderful variety of animals and plants. He invokes fellow pluralists to help his cause, but Whewell's work is summarily ignored. Pascal is summoned to highlight man's lowly and wretched place in the cosmos. Man is situated on a planet that is 'far from being the world most favourably established for the maintenance of existence. Differences of ages, positions, masses, ... biological conditions, etc. place a great number of other worlds at a degree of habitability superior to that of the earth.'[77]

Curiously, Flammarion's first convert to the cause of pluralism is Flammarion himself. The weight of his arguments in the case for extraterrestrial life is so convincing that he regards pluralism as probably innate![78] His conclusion is no less dramatic. Therein he implores his readers to join him in a declaration to God that 'we were insane to believe that there was nothing beyond the earth, and that our poor abode alone possessed the privilege of reflecting your grandeur and your power!'.[79]

Flammarion's *La Pluralité* is set firmly in the exquisite French tradition of pluralist writing. Its success owes much to the rich vein he found in the works of Cyrano de Bergerac, Fontenelle, Voltaire, and Pascal. The poetic and pleading tone is similar, as is the positive existentialism. Indeed, the book must have read like an antidote to the pessimism of the time. Many would have turned to the famous book for the fact it was written by one so young. And though its original lack of scientific detail left it open to criticism from vindictive and unforgiving scholars, it did not deter his many readers.

Indeed, the reception to *La Pluralité* was truly a mixed one. Flammarion presented a copy of his book to his guru Jean Reynaud, which, by Flammarion's own account, Reynaud 'received sympathetically, read without delay, and adopted as his own'.[80] It seems Napoleon III was also impressed, sharing the ideas of the book with the empress *et al.*, and going as far as to invite Flammarion to discuss the book with him. Alas, the astrophysicists were not so keen. Le Verrier dismissed Flammarion from the observatory.[81]

It made little difference to Flammarion. His newfound fame meant he secured a position at the Bureau des Longitudes, where he stayed until 1866. Few of Flammarion's readers were familiar with his short first edition. But thousands upon thousands read the second. The greatly expanded subsequent editions of five-hundred-plus pages showed no let-up in Flammarion's enthusiasm for extra-terrestrial life. Indeed, his pluralism remained so radical that the later editions also included reasoned sections on the possibility of lunar and solar life.

Later editions also include a fascinating section entitled 'L'Humanité dans L'Univers'. Running to more than one hundred pages, this chapter of *La Pluralité* details the ideas of planetarians such as Huygens, Kant, and Locke. Flammarion levels the charge of anthropocentrism at such authors, pointing out that their version of imagined extraterrestrial life was merely remodelled men. Instead, he takes an evolutionary approach. Varying conditions of planetary life would mean creatures different from men, and Flammarion goes on to speculate just what such creatures might look like.

The status of our lowly Earth is key. In the hierarchy of worlds beyond ours, Flammarion imagines the divine provenance and pure nature in notions such as goodness, beauty, and truth. Thus, extraterrestrials, who are also children of God, form a celestial family, and man a 'citizen of the sky'. Flammarion implies that humanity in its broadest sense is ubiquitous in the cosmos. But in a typical Flammarion twist, man's omnipresence is linked to the transmigration of souls, 'the earths which hover in space have been considered by us ... as the future regions of our immortality. There is a celestial home of many dwellings, and there ... we recognise those places which we will one day inhabit'.[82] Besides, Flammarion imagines the planets as 'studios of human work, schools where the expanding soul progressively learns and develops, assimilating gradually the knowledge to which its aspirations tend, approaching thus evermore the end of its destiny'.[83] His sense of pluralism is profound, 'Plurality of worlds; plurality of existences: these are two terms which complement and illuminate each other'.[84]

Flammarion even has rather belated advice for Bruno and Galileo. Religion and pluralism he admits are unhappy bedfellows. That does not mean, however, that Flammarion is short of ideas. He proposes no less than four solutions to the challenges pluralism poses for Christianity: (i) God became concurrently incarnate and died on all those planets where there was sin; (ii) God became serially incarnate, on different planets, at different times; (iii) God visited only Earth, as only here did sin occur; and (iv) Christ's earthly deeds brought cosmic redemption to all worlds, the last being favoured by Flammarion himself.

Flammarion: alien fiction in spacetime

Many of Flammarion's other projects also proved to be durable. He founded the Societé Astronomique de France, and a Flammarion publishing house, both of which endure. He also established a telescopic observatory, twenty miles south of Paris at the Juvisy-sur-Orge estate gifted to Flammarion by an admirer of his work, which is still operational. Flammarion's other notable work on pluralism was a speculative mélange of science and fiction, *Les Mondes Imaginaires et Les Mondes Réels (Real and Imaginary Worlds)* published in 1864. The book was a meticulous history and critical survey of explorations in the idea of extraterrestrial life, both in philosophy and in the cosmic voyage genre. In its pages Flammarion admits the inspiration of Kepler, Godwin, Cyrano de Bergerac, and Jonathan Swift in his fictional work on the habitability of heavenly bodies.

Flammarion was an ardent communicator of science. He was possessed by a passion to engage the mass of ordinary people with science, using various narrative techniques to further his cause. It was only a matter of time before his fertile imagination used science fiction, as well as science. In so doing, Flammarion was a great pioneer. Like H. G. Wells in England, he cultivated the public's taste for cosmic science. His 1872 *Recits de L'Infini*, translated as *Stories of Infinity* in 1873, included three fascinating and richly imaginative stories of life in the universe. His tales *Infinity*, *The History of a Comet*, and particularly *Lumen*, tell of a disembodied travelling spirit, which observes the potential range of wondrously exotic cosmic life.

Lumen is a testament to the way in which an unbridled science-fictional imagination can support the science of thought-experimentation. Written before the literary style of modern science fiction had been more fully developed by the likes of Wells, *Lumen* is based on the dialogic style of science communication, particularly Humphry Davy's *Consolations in Travel; or, The last Days of a Philosopher* (1830) and the much earlier *Entretiens* of Fontenelle. In *Lumen*, Flammarion was the first author to thoroughly apply the theory of evolution, albeit Lamarckian, to the creation of truly alien life-forms, thereby laying one of the fundamental keystones to twentieth-century science fiction:

> We have grown so used to the idea of alien beings since H. G. Wells found a melodramatic role for them to play in *The War of the Worlds* that it is hard to imagine a time when the idea was new and wonderfully exotic.[85]

Kepler may warrant credit for inventing the alien in *Somnium*. But Flammarion was the first writer to 'extrapolate that notion to its hypothetical limit'[86] and to 'fill that range with examples by the dozen'.[87] As we shall soon see, Wells' masterstroke was to reflect the inhumanity of the alien back on humanity itself, 'To me it is quite credible that the Martians may be descended from beings not unlike ourselves, by a gradual development of brains and hands ... at the expense of the rest of the body'.[88] In the immediate future it was the Wellsian image of the alien as monstrous that would prevail. Only later did the Flammarionesque idea of alien life as a precious fifth element reappear.

Not content with having created a benign alien in *Lumen*, Flammarion went on to hint at the nature of spacetime. Thirty years before Einstein's Theory of Special Relativity, *Lumen* was the earliest science fiction novel to proffer the principle that time and space are not absolute. They exist, Flammarion suggested, only relative to one another. He goes on to picture the changes in observation that may result from travelling at velocities close to, and beyond, the speed of light, considering that faster than light travel would render history in reverse. Indeed, Flammarion's very notion of space as a seething sea of 'undulations', replete with latent and pent-up energies, is equally inspired.

H. G. Wells on time and space

H. G. Wells was not the first Victorian author to harness the machine to explore Darwinian and alien notions of spacetime. In Bulwer-Lytton's *The Coming Race*, the harnessing of technology had made the Vril-ya evolutionarily superior. And in 1872, Samuel Butler's irreverent satire, *Erewhon*, tells the tale of a traveller who uncovers the beautiful faraway realm of Erewhon, but soon learns this idyll has its flaws.

The most Darwinian part of *Erewhon* is his 'Book of the Machines'. It revolves around the question of whether the machine is servant of man, or man servant to the machine. Butler's purpose was to mock those who argued that the world had been designed for man. He objected to the contradictory concept of evolution as mechanical and at the same time unsolicited. If evolution is the result of accidental changes mechanically perpetuated, then it is best left to describing the progress of machines. A theory that regards man as a machine is no better than an absurdity that supposes machines to be animate.

With ironic chance insight, Butler had invented machine consciousness. *Erewhon* is the prototype of a twentieth-century obsession. Butler can never have dreamt of the thousands of artificial intelligences that would follow. Ever since Butler, science fiction has had double vision on the metaphor of the machine. At times, machines mediate between man and the cosmos, human and nonhuman, acting as the agency of man's protection. On other occasions, machines act as a medium for the nonhuman, a threat to the human condition. Witness H.G. Wells' *The Time Machine* (1895) and *The War of the Worlds* (1898), where the metaphor flows both ways, and Wells masters the alien and machine both as symbols for the power of reason, and as a diabolical mechanism.

Herbert George Wells was spawned from an English lower middle class that had previously generated another key author in Charles Dickens. Wells' mother had been in service, his father a gardener. Though optimistic of becoming socially mobile as shopkeepers, the shop failed, year after year. Wells' own employment began as apprentice to a draper. But it ended rather sharply when he was told he was not refined enough to be a draper. Such rejection at the sharp end of a class-conscious society became one of the key influences on Wells' critique of the Victorian world.

Wells' defining moment in science came on meeting 'Darwin's Bulldog', the great T.H. Huxley. Wells was bright enough to have won a scholarship to the Normal School of Science, later the Royal College of Science, where he studied evolutionary biology under Huxley. A fervent Darwinian, Huxley was the chief science communicator in England at the time. He had introduced the word 'agnostic' into the lexicon, and impressed man's simian ancestry on the public imagination. His public lectures attracted massive audiences. Two thousand were reportedly turned away at St Martin's Hall in 1866, the year of Wells' birth.[89]

Huxley's crowning triumph had come from the infamous conflict with Bishop Wilberforce. On 30 June 1860, in front of a huge audience at a now legendary meeting of the British Association for the Advancement of Science in Oxford, Wilberforce had asked whether it was through Huxley's grandfather or his grandmother that he claimed descent from a monkey. Huxley's reply was exquisite:

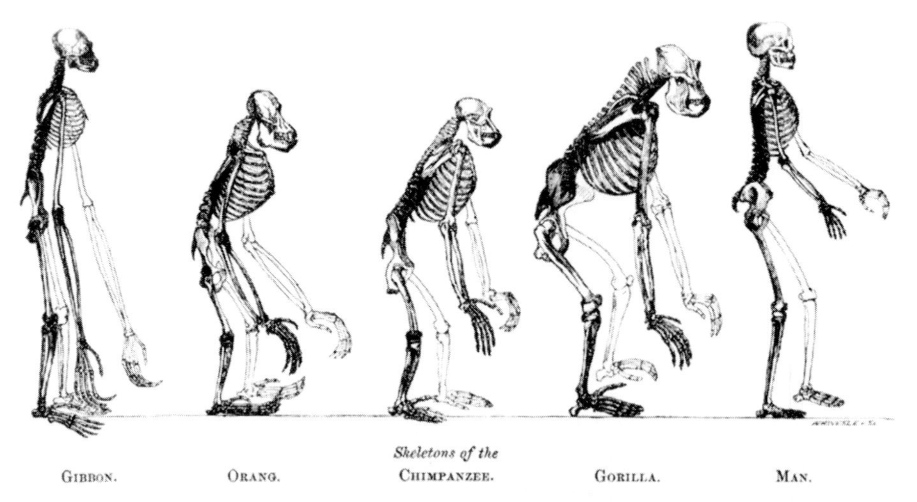

Figure 14 The frontispiece from Huxley's 1863 *Evidence as to Man's Place in Nature*, the first appearance of the famous image comparing the skeletons of apes and Man.

> A man has no reason to be ashamed of having an ape for his grandfather. If there were an ancestor whom I should feel shame in recalling it would rather be a man who plunges into scientific questions with which he has no real acquaintance, only to obscure them by an aimless rhetoric.[90]

The attack on religious prejudice had electrified the meeting, and Huxley had been the hero of the day. The feud finally ended when Wilberforce dashed his head while out horse riding, making him fortunately oblivious to Huxley's last judgment, 'for once reality and his brain came into contact, and the result was fatal'.[91]

With Huxley as his inspiration, Wells began as a writer, living in the dark, lanterned, black macadam streets of Victorian London, engine-room of the British Empire. The first of Wells' seminal novels, *The Time Machine*, had plotted a dark future for man. The book was a sceptical view of the devilish enginery of progress and imperialism.

It was an instant triumph. *The Time Machine* revealed a dread of proletarian uprising in its antipathy towards the Morlocks, the troglodyte species that dwells underground in the book. But Wells eschewed a contemporary Marxist emphasis on class hostility by changing class difference into species difference.[92] Indeed, the mood of the age was one of fear. Between 1870 and the start of World War I there were hundreds of authors writing invasion literature, often topping the bestseller lists in many countries of Europe, and the United States. This pervading sense of *fin de siècle*, and *fin de globe*, was fed limitless power by Darwinian evolution. It allowed Wells to produce work whose appeal has dimmed little since.

And so a host of spectres had haunted the age. Into this fragile climate marched an alien invasion. In perhaps the most superb opening in the entire history of science fiction, Wells' *The War of the Worlds* (1898) does for space what his *The Time Machine* did for time:

> No one would have believed in the last years of the nineteenth century that this world was being watched keenly and closely by intelligences greater than man's and yet as mortal as his own ... With infinite complacency men went to and fro over this globe about their little affairs, serene in their assurance of their empire over matter.[93]

Wells' Martians are agents of the void. In Jules Verne's *Journey to the Centre of the Earth*, man had penetrated space, in the form of the unexplored terrestrial interior. Now, in *War of the Worlds*, it is space that powerfully shatters the human domain. Wells' book is the finest and most influential of all alien contact narratives. It is the first Darwinian fable on a universal scale.

Wells' wrath was directed at the idea of the 'becoming of man'. The book begins with a quote from Kepler, 'But who shall dwell in these worlds if they be inhabited? ... Are we, or they, Lords of the World? ... And how are all things made for man?'.[94] It is a battle between Earth and Mars, humans and Martians. The narrator of this struggle for survival is a philosopher, writing a thesis on the progression of moral ideas with civilisation. His assumption of a bright future is rudely blown apart in mid-sentence by the brutal natural force of evolution in the shape of the Martian attack.

The War of the Worlds features the first 'menace from space'. Regardless of Kepler's lunar serpents and Swift's malevolent Laputians, the modern alien owes everything to Wells. With their distinctive physiology and intellect, the book's Martians are the archetypal alien. Wells' story was first serialised in 1897, the same year that Kurd Lasswitz, father of German science fiction, published his contrasting view of peace-loving Martians in *Auf Zwei Planeten* (*On Two Planets*).

One reason that the books of Wells and Lasswitz met their success was down to difference. Other space novels at the time, especially those about Mars, spoke of travellers journeying through the Solar System on some kind of imperial and Victorian conquest. The denouement of their meeting with extraterrestrials was invariably a sense of triumphant superiority, garnered with an enhanced human self-esteem. In Wells and Lasswitz, however, it is the Martians who conquer space. They land on Earth and pose a serious threat. The fear is real and gripping.

Figure 15 The Martian Tripods illustrated by Alvim Corréa, 1906.

Wells had been keenly aware of the possibility of extraterrestrial life for some time. He had contributed to the pluralist debate in 1888 at the Royal College of Science on the subject 'Are the Planets Habitable?'. He had also written essays in support of fellow pluralists Kepler, Camille Flammarion, and Percival Lowell, whose mistaken obsession with Schiaparelli's 'evidence' of life on Mars had recently reached Europe. Rather than being a whimsical work of fiction, *The War of the Worlds* demolishes the idea that man is the peak of evolution. Instead Wells creates the myth of a technologically superior alien intelligence.

Mars is a dying world. Its seas are evaporating, its atmosphere melting away. The entire planet is doomed so, 'to carry warfare sunward is, indeed, their only escape from the destruction that generation after generation creeps upon them'.[95] Thus the terror of the void is brought down to Earth. Wells delivers repeated reminders of 'the immensity of vacancy in which the dust of the material universe swims'[96] and evokes the 'unfathomable darkness'[97] of space. Life is portrayed as precious and frail in a cosmos that is essentially deserted.[98]

The book consciously conveys the quality of the void – immenseness, coldness, and indifference – in its rendering of the aliens. But it is the Martian machines that vividly hammer home the cosmic chain of command:

> It is remarkable that the long leverages of their machines are in most cases actuated by a sort of sham musculature ... Such quasi-muscles abounded in the crablike handling-machine ... It seemed infinitely more alive than the actual Martians lying beyond it in the sunset light, panting, stirring ineffectual tentacles, and moving feebly after their vast journey across space.[99]

The Martian Tripods tower over men in the flesh, as the vast intellects of their occupants tower over human intelligence. Physically frail, but mentally intense, the Martians and their superior machines are instruments of human oppression. Their weapons of heat rays and poison gas are dehumanising devices of mass murder. All efforts at contact are futile, furthering the idea of the aliens as an unrelenting force of the void.

Understandably, the human reaction to this cosmic struggle is alienation.[100] The narrator alienates himself from the grim reality of the inevitable triumph of death over life: 'I suffer from the strangest sense of detachment from myself and the world about me; I seem to watch it all from the outside, from somewhere inconceivably remote, out of time, out of space, out of the stress and tragedy of it all'.[101]

Wells planned the invasion on bicycle. It is pleasing to picture him mapping mayhem as he 'wheeled about the district marking down suitable places and people for destruction by my Martians'.[102] As early as 1896 he declared his intentions to 'completely destroy Woking – killing my neighbours in painful and eccentric ways – then proceed via Kingston and Richmond to London, which I sack, selecting South Kensington for feats of particular atrocity'.[103] It is this exquisite violence of Wells' imagination that marks his genius.[104]

And, as Wells had planned on bicycle, so it was in fictional South Kensington. The area is haunted by the sound of the Martians howling, 'Ulla, ulla, ulla, ulla'.[105] The alien invaders have finally fallen prey to an earthly predator – microbes. The fate of the Martians emphasises not only the insignificance of human resistance to the struggle, but also the latent power of unsolicited natural selection.

Wells' seminal alien story ends with the dialectic of the Martians. On one hand, the narrator feels for the aliens. The tragedy of the Martians is the tragedy of man.[106] The aliens' torment speaks strongly of the common struggle for all life in a hostile universe. On the other hand, Wells has constantly portrayed the Martians as vast, cool, and unsympathetic. Alien in tooth and claw, as it were.

The narrator may merely be projecting emotion onto creatures that are fundamentally inhuman.

Wells' Martians are a fascination. They are alien, yet they are human. They are what we may one day become, with their 'hypertrophied brains and atrophied bodies',[98] the tyranny of intellect alone. But they are also political. Wells evidently has the Martians brutally colonise Earth, but, 'before we judge them too harshly we must remember what ruthless and utter destruction our own species has wrought . . . upon its own inferior races . . . Are we such apostles of mercy as to complain if the Martians warred in the same spirit?'.[107]

Finally, in writing of the alien, Wells is also writing about his own Victorian machine world. The Martians are a veiled criticism of the machine age,[108] with its application of science to industry. In this fashion, *The War of the Worlds* condemns the social machine of capitalism, the reducing of humans to anonymous cattle, the indifference at any attempts to communicate the inhumanity of the system, and the feeling of alienation.

And it is a Wellsian notion of time as the fourth dimension that we shall find in the next chapter, 'Einstein's sky'. It is the fabric of a universe of effortlessly wheeling galaxies in gently curving spacetime. Into this bible-black sky, pluralism now plunged, trading the Wellsian terror of the void for a Flammarion freedom of infinite spacetime. 'All philosophy', Fontenelle had written, 'is based on only two things: curiosity and poor eyesight; if you had better eyesight you could see perfectly well whether or not these stars are solar systems.'[109] In the early twentieth century, curiosity and the outward urge took pluralism to infinity, and beyond.

Notes

1 Flammarion, C. (1872) *Lumen*, trans. B. Stableford (2002), Connecticut
2 Wells, H. G. (1898) *The War of the Worlds*, Gollancz edition, SF Masterworks (2003)
3 Brake, M. and Hook, N. (2007) *Different Engines: How Science Drives Fiction and Fiction Drives Science*, London
4 Lovejoy, A. (1936) *The Great Chain of Being*, Cambridge, Mass.
5 *De animalibus historia*, quoted in Lovejoy, A. (1936) *The Great Chain of Being*, Cambridge, Mass.
6 St Thomas Aquinas, *Summa contra Gentiles*
7 Lovejoy, A. (1936) *The Great Chain of Being*, Cambridge, Mass.
8 Montaigne, M. de (1595) *The Complete Essays*, trans. M. A. Screech (1987) London
9 Lovejoy, A. (1936) *The Great Chain of Being*, Cambridge, Mass.
10 Eisley, L. (1970) *The Firmament of Time*, New York
11 Lovejoy, A. (1936) *The Great Chain of Being*, Cambridge, Mass.
12 Ruse, M. (1979) *The Darwinian Revolution*, Chicago
13 Huggins, W. (1909) *The Scientific Papers of Sir William Huggins*, ed. Sir William Huggins and Lady Huggins, London, p. 60

14 Davies, R. (2008). *The Darwin Conspiracy: Origins of a Scientific Crime*, London, p. 164

15 Darwin, F., ed. (1888) *The Life and Letters of Charles Darwin: Including an Autobiographical Chapter*, London, 3:18

16 Darwin, C. (1871), *The Descent of Man, and Selection in Relation to Sex*, 1st edn., London

17 Fichman, M. (2004) *An Elusive Victorian: The Evolution of Alfred Russel Wallace*, Chicago

18 ibid.

19 Wallace, A. R. (1903) Man's Place in the Universe, *Fortnightly Review* **73**, 1 March, pp. 395–411

20 ibid., p. 411

21 ibid., p. 396

22 ibid., p. 411

23 Wallace, A. R. (1904) *Man's Place in the Universe: A Study of the Results of Scientific Research in Relation to the Unity or Plurality of Worlds*, 4th edn., London, pp. 326–36

24 ibid., pp. 326–7

25 ibid., p. 335

26 Quoted in Crowe, M. J. (2008) *The Extraterrestrial Life Debate, Antiquity to 1915, A Source Book*, Notre Dame, p. 437

27 ibid., p. 429

28 Sagan, C. (1973) *Hypothesis*, in Ray Bradbury *et al.*, *Mars and the Mind of Man*, New York, p. 15

29 Quoted in Crowe, M. J. (2008) *The Extraterrestrial Life Debate, Antiquity to 1915, A Source Book*, Notre Dame, p. 384

30 As quoted in Willard, C. R. (1894) Richard A. Proctor, in *Popular Astronomy*, **1**, p. 319

31 Proctor, R. A. (1882) *The Borderland of Science*, London, p. v.

32 Proctor, R. A. (1880) Other Worlds and Other Universes, in *Myths and Marvels of Astronomy*, new edn., London, p. 135

33 Fraser, J. (1873) Proctor the Astronomer, in *English Mechanic*, **18**, p. 322

34 Proctor, R. A. (1870) *Other Worlds Than Ours*, London, p. 4

35 ibid., p. 141

36 ibid., p. 145

37 As quoted in Townsend, L. T. (1914) *The Stars Not Inhabited*, New York, pp. 143–5

38 Proctor, R. A. (1870) *Other Worlds Than Ours*, London, pp. 84–5

39 Proctor, R. A. (1873) *The Borderland of Science*, London, pp. 156–7

40 ibid.

41 Proctor, R. A. (1875) *The Science Byways*, London, p. 35

42 ibid., pp. 34–5

43 Proctor, R. A. (1876) *Our Place Among Infinities*, London, p. 67

44 Proctor, R. A. (1875) *The Science Byways*, London, p. 22

45 ibid., pp. 27–8

46 ibid.

47 Proctor, R. A. (1875) *The Science Byways*, London, p. 68

48 Proctor, R. A. (1876) *Our Place Among Infinities*, London, pp. 69–70

49 Anonymous (1888) Richard A. Proctor Dead, *New York Times*, 13 September 1888, p. 1

50 Proctor, R. A. (1877) *Myths and Marvels of Astronomy*, New York, p. 109

51 Proctor, R. A. (1881) *The Poetry of Astronomy*, London, pp. 180–1

52 Avertissement de l'editeur, prefixed to the 1866 edition of *Voyages et aventures du capitaine Hatteras*
53 Rose, M. (1982) *Alien Encounters: Anatomy of Science Fiction*, Harvard
54 ibid.
55 Bulwer-Lytton, E. (1871) *The Coming Race* (in *Bulwer's Novels and Romances*, 1897) New York
56 Sherard, R. A. (1894) Flammarion the Astronomer, *McClure's*, **2**, pp. 569–77
57 Houzeau, C. and Lancaster, A. (1882, reprint 1964) *Bibliographie générale de l'astronomie ... jusqu'en 1880*, London, 2, lxxiv
58 As quoted in Sherard, R. A. (1894) Flammarion the Astronomer, *McClure's*, **2**, p. 569
59 Newcomb, S. (1894) A Very Popular Astronomer, *Nation*, **59**, 469–70: 469
60 Proctor, R. A. (1881) *The Poetry of Astronomy*, London, pp. 250–4
61 Flammarion, C. (1890) How I Became An Astronomer, *North American Review*, **150**, 100–5: 102
62 ibid., p. 103
63 Quoted in Crowe, M. J. (2008) *The Extraterrestrial Life Debate, Antiquity to 1915, A Source Book*, Notre Dame, p. 379
64 Flammarion, C. (1882) *Les Mondes Imaginaires et Les Mondes Réels*, 20th edn., Paris, p. 566
65 Flammarion, C. (1911) *Mémoires*, Paris, p. 202
66 ibid., pp. 203–4
67 ibid.
68 Flammarion, C. (1862) *La Pluralité Des Mondes Habités*, Paris, p. 7
69 ibid., p. 8
70 Quoted in Crowe, M. J. (2008) *The Extraterrestrial Life Debate, Antiquity to 1915, A Source Book*, Notre Dame, p. 409
71 ibid.
72 ibid., p. 33
73 ibid., p. 25
74 ibid., p. 26
75 ibid., p. 43
76 ibid., p. 37
77 ibid., p. 50
78 ibid., pp. 7, 45
79 ibid.
80 Flammarion, C. (1911) *Mémoires*, Paris, p. 242
81 Crowe, M. J. (2008) *The Extraterrestrial Life Debate, Antiquity to 1915, A Source Book*, Notre Dame
82 Flammarion, C. (1885) *La Pluralité Des Mondes Habités*, 33rd edn., Paris, p. 320
83 ibid., p. 328
84 ibid., p. 324
85 Brake, M. and Hook, N. (2007) *Different Engines: How Science Drives Fiction and Fiction Drives Science*, London
86 ibid.
87 ibid.
88 Wells, H. G. (1898) *The War of the Worlds*, London
89 Carey, J. (1995) *The Faber Book of Science*, London
90 Huxley, J. (1903) *Life and Letters of Thomas Henry Huxley*, London

91 ibid.
92 Huntington, J. (1995) *The Time Machine* and Wells' Social Trajectory, *Foundation*, 65, Liverpool
93 Wells, H. G. (1898) *The War of the Worlds*, Gollancz edition, SF Masterworks (2003)
94 ibid.
95 ibid.
96 ibid.
97 ibid.
98 Rose, M. (1982) *Alien Encounters: Anatomy of Science Fiction*, Harvard
99 Wells, H. G. (1898) *The War of the Worlds*, Gollancz edition, SF Masterworks (2003)
100 Philmus, R. M. (1970) *Into the Unknown*, Los Angeles
101 Wells, H. G. (1898) *The War of the Worlds*, Gollancz edition, SF Masterworks (2003)
102 Wells, H. G. (1934) *Experiment in Autobiography*, New York
103 ibid.
104 Huntington, J. (1999) My Martians: Wells' Success, *Foundation*, 77, Liverpool
105 Wells, H. G. (1898) *The War of the Worlds*, Gollancz edition, SF Masterworks (2003)
106 Huntington, J. (1999) My Martians: Wells' Success, *Foundation*, 77, Liverpool
107 Wells, H. G. (1898) *The War of the Worlds*, Gollancz edition, SF Masterworks (2003)
108 Rose, M. (1982) *Alien Encounters: Anatomy of Science Fiction*, Harvard
109 Flammarion, C. (1872) *Lumen*, trans. B. Stableford (2002) Connecticut

6

Einstein's sky: life in the new universe

. . . we regard the cosmos as very beautiful. Yet it is also very terrible. For ourselves, it is easy to look forward with equanimity to our end, and even to the end of our admired community; for what we prize most is the excellent beauty of the cosmos. But there are the myriads of spirits who have never entered into that vision. They have suffered, and they were not permitted that consolation. There are, first, the incalculable hosts of lowly creatures scattered over all the ages in all the minded worlds. Theirs was only a dream life, and their misery not often poignant; but none the less they are to be pitied for having missed the more poignant experience in which alone spirit can find fulfilment. Then there are the intelligent beings, human and otherwise; the many minded worlds throughout the galaxies, that have struggled into cognizance, striven for they knew not what, tasted brief delights and lived in the shadow of pain and death, until at last their life has been crushed out by careless fate. In our solar system there are the Martians, insanely and miserably obsessed; the native Venerians, imprisoned in their ocean and murdered for man's sake; and all the hosts of the forerunning human species. A few individuals no doubt in every period, and many in certain favoured races, have lived on the whole happily. And a few have even known something of the supreme beatitude. But for most, until our modern epoch, thwarting has outweighed fulfilment; and if actual grief has not preponderated over joy, it is because, mercifully, the fulfilment that is wholly missed cannot be conceived.

Olaf Stapledon, *Last and First Men*[1]

Our position in the material universe is special and probably unique, and . . . it is such as to lend support to the view, held by many great thinkers and writers today, that the supreme end and purpose of this vast universe was the production and development of the living soul in the perishable body of a man.

Alfred Russel Wallace, *Man's Place in the Universe, as Indicated by the New Astronomy*[2]

> The sustaining motive of our pilgrimage had been the hunger which formerly drove men on Earth in search of God. Yes, we had one and all left our native planets in order to discover whether, regarding the cosmos as a whole, the spirit which we all in our hearts obscurely knew and haltingly prized, the spirit which on Earth we sometimes call humane, was Lord of the Universe, or outlaw; almighty, or crucified. And now it was becoming clear to us that if the cosmos had any lord at all, he was not that spirit but some other, whose purpose in creating the endless fountain of worlds was not fatherly toward the beings that he had made, but alien, inhuman, dark.
>
> Olaf Stapledon, *Star Maker*[3]

Man's place in the cosmos

Earth was an alien planet. Indeed, by the dawn of Einstein's century, Earth had been alien for some time. The paradigm shift of the Copernican revolution had cut both ways. Not only had the revolution made Earths of the planets, it had also brought the alien down to Earth. The old universe of Aristotle had been small, static, and Earth-centred. It had the stamp of humanity about it. Constellations boasted the names of earthly myths and legends, and a magnificence that would later gift the Church sufficient validation of God's glory.

The new universe was inhuman. The deeper the telescopes probed, the darker and more alien it became. 'The history of astronomy', suggests novelist Martin Amis, 'is a history of increasing humiliation. First the geocentric universe, then the heliocentric universe. Then the eccentric universe – the one we're living in. Every century we get smaller. Kant figured it all out, sitting in his armchair ... The principle of terrestrial mediocrity.'[4]

Carl Sagan, the American astronomer and science fiction author, had gone further.[5] Sagan suggested that humans had suffered a series of 'Great Demotions' in the last five centuries. First, consider Earth. It was not at the centre of the universe. Nor was it the only object of its kind, made of a unique material only to be found on terra firma. Next, consider the Sun. Not at the centre of the universe, not the only star with planets, nor eternal. If cosmology could be endured, there was biology. Man now lived among the microbes. He had no special immunity from natural law, and there was vanishingly little evidence of a divine image. Each successive demotion had a huge impact, both on the human condition, and on the meaning of life in the cosmos.

Then, there was Einstein's sky. As the twentieth century unfolded, astronomers would uncover a cosmos composed of more stars than grains of sand on all the beaches of planet Earth. The infant century was about to boast a vast Milky Way Galaxy, not at the centre of the cosmos, nor the only galaxy within it. A hundred billion other galaxies would be discovered; adrift in an expanding universe so immense that light from its outer limits takes longer than twice the age of the Earth to reach terrestrial telescopes. And there may be other universes.

And so this chapter considers the impact of Einstein's sky on the pluralism of the early twentieth century. We shall be mindful, as the story unfolds, of the words of Darwin's Bulldog, English biologist T. H. Huxley, 'the question of questions for mankind – the problem which underlies all others, and is more deeply interesting than any other is ascertainment of the place which Man occupies in nature and of his relations to the universe of things'.[6] And mindful too of the view of Carl Sagan, who believed the final demotion would be the discovery of another biological intelligence in the universe. We shall see, as the centuries drew on, that the stage upon which the debate played out grew larger and larger. And so, the journey thus far travelled.

Pluralism and anthropocentrism

So, to begin at the beginning: the geocentric universe that was to emerge as the dominant cosmology of the ancient Greek world. Even meticulous observations of the night sky by our ancestors led to a misplaced sense of the importance of man. Plato's idealist perspective was one of a divine cosmos, a universe in which change was cast out, as no doubt Plato and Aristotle would have cast it out of society. The crisis that faced the ancient Greek world bled into its physics, as the philosophers of the day tried to save society from the sociopolitical disaster into which it was descending.

So it was on Earth, so it shall be in the heavens. In Plato's vision, the philosopher's highest calling became the consideration of perpetuity. In fear of change, most Greek scholars believed the divine nature of the cosmos shone through in terms of the unchanging and regular paths of the planets, with orbits of perfect and circular movement. In Aristotle's vision, the universe was geocentric. But for him our world was not just a physical centre. It was also the centre of motion. Everything in the universe moved with respect to this single centre. Aristotle said that if there were more than one world, more than just a single centre, elements such as earth would have more than one natural place to fall to. It was a rational and natural contradiction. Aristotle declared the Earth unique, and man alone in the universe.

The medieval sky too was Aristotelian, and geocentric. But it was now a vision utterly transformed by the Church. Through Christian symbolism, Aristotle's universe of spheres mirrored man's hope and fate. Both bodily and spiritually, man sat midway. His crucial position in the hierarchical Chain of Being was halfway between the inert clay of the Earth's core and the divine spirit of the heavens. And though the only known and observed life in the cosmos was here on Earth, in filth and uncertainty, close to Hell, man was always under the eye of an assumed God. The choice was whether to follow a human and earthly nature down to its natural place in Hell, or to engage with the spirit, and follow the soul up through the celestial spheres to God. Any alteration in the great plan of Aristotle's universe was bound to corrupt this drama of Christian life and death. To shift the Earth was to sever the continuous chain of created being, and to move the Throne of God himself.

Nonetheless, shift it we did.

Heliocentrism was born in the Copernican revolution, and with it a new tradition in the plurality of worlds. The Copernican worldview struck at anthropocentrism: if the Earth is a planet, then the planets may be Earths; if the Earth is not central, then neither are we.[7] Copernicus himself was silent on pluralism. But there was a new sense of wonder at the emerging possibilities. Giordano Bruno paid the price for imagining his universe filled with inhabited worlds. And this same sense of wonder was responsible for the birth of science fiction.

So science fiction too began with the revolution of Copernicus. It was born along with the paradigm shift of the old universe into the new. Aristotle's cosy geocentric cosmos was anthropocentric. The new universe of Kepler and Galileo was, at least potentially, decentralised, infinite, and alien. In many ways, the early space fiction can be read as a response to the cultural shock created by the discovery of humanity's marginal position in a universe fundamentally inhospitable to man.

Pluralist science fiction embraced the alien. It emerged as our attempt to make human sense of Copernicus' new universe. Witness Kepler's invention of the alien in *Somnium*. The extraterrestrials that stalk *Somnium*'s Moon world are creatures fit to survive an alien habitat. It was the first space fiction of the age. But the alien voyages evolved quickly, and became a potent way for exploring the insignificance of man.

Cyrano de Bergerac's *L'autre Monde* (*Other Worlds*) trilogy was among the first to use satire in its portrayal of man's place in the cosmos. In a series of weird events on the Moon and Sun, Cyrano addresses anthropocentrism head-on by exposing the inferiority of humans in an alien setting. On the Moon, his anti-hero stands trial for believing that his own planet is a world, and not a 'moon'. The Lunarians believe in the total insignificance of humans, as their human counterparts might do on Earth. Indeed, the bi-peds are thought to be monsters devoid of reason, and are best described as 'plucked parrots'.

Travels to the Sun told a similar tale. Man is portrayed as arrogant, a deluded creature who believes the entire animal world, and the environment, to be at his disposal. But the solarian birds are equally arrogant. They too believe themselves the most supremely rational and cultured beings in the cosmos. In this way, Cyrano satirised the bigotry and beliefs of the age. *L'autre Monde* was a secular picture of the potential meaning of a universe fit for life. It was a Copernican stand against the anthropocentric notion that man was the centre of creation.

Cyrano's countryman Voltaire also used satirical space fiction to mock the geocentric conceit. He portrays man as midway between telescope and microscope, but makes it clear that his visitors to Earth in *Micromégas* are scathing about man's arrogant anthropocentrism. Indeed, the very discovery of the aliens, including Micromégas himself, demotes the idea of man as the crowning of Creation. Man is not the measure of all things, says Voltaire. Man boasts conclusive proof on the nature of the universe, in face of the fact there are as many theories as thinkers.

And so geocentrism gave way to heliocentrism, which yielded eventually to galactocentrism. Just a couple of decades after Voltaire had penned *Micromégas*, William Herschel was sweeping the night sky with state-of-the-art handmade telescopes of the day. In mid century, Thomas Wright and Immanuel Kant had proposed the idea that the fuzzy patches of light, known as nebulae, were actually distant 'island universes' composed of myriad star systems. The form of our Milky Way Galaxy was expected to be similar to the nebulae.

By 1783, two years after he had discovered Uranus and doubled the size of the Solar System, Herschel began his attempts to gauge the size and scale of the Galaxy by an analysis of stellar magnitude and position. In an extrapolated geocentrism, Herschel came to the conclusion that the Milky Way encircled the Earth, and that, like some of the nebulae in deep space, the Galaxy was the shape of a flattened disc. The conclusion was clear. The Earth, or at least the Sun about which it moved, was at or near the centre of the Galaxy. Though Herschel's methodology was flawed, his model remained relatively unscathed for the next hundred years, given the odd minor tweak.

Paradoxically, even though he was a galactocentrist, by dwelling on the very construction of the cosmos, Herschel struck out against anthropocentrism. A prophet of deep space astronomy, while others peered at the familiar local planets, Herschel was probing the depths of distance and the unidentified. While professional astronomers were tinkering with planetary distances in the Solar System, Herschel the amateur was charting star systems beyond the professional imagination. And while they were using the rough proportions of the speed of light to infatuate over the mechanics of the Jovian system,

Herschel realised he was gazing so far into deep space that he was looking millions of years into the past.

The biological theory of evolution also had an effect on anthropocentrism. Though its immediate impact on the place of man among terrestrial animals was quite profound, its influence on the question of extraterrestrialism was more gradual. But as evolutionary theory threatened the image of man due to his descent from lowly creatures, the even more all-embracing idea of pluralism promised a worldview in which man was but one link in a cosmic chain of life that stretched far into deep space. By the beginning of Einstein's century this biological cosmos was a recognisable worldview. It informed the fictions of Flammarion and H.G. Wells, among others, and was to influence the narratives of one of Wells' successors Olaf Stapledon, as we shall see.

During the course of the twentieth century, the tension between anthropocentrism and pluralism grew. As Einstein's influence diffused into the scientific culture, the early decades saw radical changes in the astronomical worldview. Whereas it might still have been possible in 1903 to argue an anthropocentric cosmos, by 1930 evidence had all but annihilated that argument. It was as Einstein had predicted: an expanding universe. But astronomers had also uncovered a cosmos of enormous dimensions. The Solar System seemed washed up on the shore of but one galaxy among millions. And a tipping point was passed after which the belief in other worlds became commonplace. The 'assumption of mediocrity', or the 'principle of terrestrial mediocrity' as Amis suggested Kant had put it, became an integral belief in standard cosmologies.

The sciences began to merge as the century dawned. Both the physical and biological elements of the cosmos needed to be addressed to avoid future anthropocentric claims on our place in the universe. Those elements of pluralism, the physical and the biological, became increasingly testable, as the decades unfolded. And yet, of course, the materialist or idealist assumptions, which lay beneath the interpretation of such scientific observations, remained. As the century blinked into existence, a new champion of anthropocentrism emerged in Alfred Russel Wallace. His position is not only instructive in and of itself, but is also indicative of a position drawn from the fledging century's pre-Einstein sky.

Alfred Russel Wallace and the privileged position of man

As an evolutionary biologist it was understandable that Wallace should speculate on the position of man in the cosmos. But he went one step further than T. H. Huxley before him. Huxley had penned his book *Man's Place in Nature* in 1863. Now, early in the twentieth century, Wallace was to take the argument from Earth into deep space. Wallace's epiphany on recent advances in astronomy had come when he was writing his book *The Wonderful Century*, in 1898. It seems Wallace was influenced in his anthropocentric view of the universe by recent galactocentric arguments. To his amazement, Wallace read the view of John Herschel and Sir Norman Lockyer that our Sun sat at the centre of the Milky Way. Further research implied that our Sun, along with its system, sat at the centre of the entire cosmos. The profound meaning of this privileged position, along with recent findings that suggested Earth was the only inhabited planet in our Solar System, led Wallace to wonder if indeed Earth was the only life-bearing planet in the entire cosmos:

> For many years I had paid special attention to the problem of the measurement of geological time, and also that of the mild climates and generally uniform conditions that had prevailed throughout all geological epochs; and on considering the number of concurrent causes and the delicate balance of conditions required to maintain such uniformity, I became still more convinced that the evidence was exceedingly strong against the probability or possibility of any other planet being inhabited.[8]

Galactocentrism sat at the heart of Wallace's worldview. The Sun was situated at the centre of a finite stellar universe. And it was the only position in this starry cosmos that was suitable for life. Thus, Wallace concluded that man's position in the material universe was special. Indeed, galactocentrism seems to have been the major physical spur in Wallace's belief of man's unique position, and that the very reason for the existence of the cosmos itself was the creation and evolution of the living soul.

So, taking cutting-edge astronomy as his lead, Wallace's major thesis for Earth's unique position was founded on a triad of necessary conditions. First, that life can exist only around our Sun, or the cluster of suns in the solar neighbourhood. Second, that no life exists on planets around other suns in the solar neighbourhood. And third, that no life exists in our Solar System beyond planet Earth. And so the assumed position of the Sun in the cosmos became Wallace's rationale for anthropocentrism. With some justification, he argued that his view was the consensus of the most renowned astronomers of the day. This much was true. When Wallace's book went to press, not only all stars, but also all observable bodies in the known cosmos, were thought to be part of a single system.

Indeed, the Milky Way *was* the cosmos. Or so they believed. And this Galaxy-as-universe model, a system bound by gravity, was something of a return to the cosy, geocentric cosmos of the past. Unlike today's Galaxy of one hundred thousand light years in diameter in an expanding Einsteinian sky, *this* Milky Way of Wallace's was barely a few light years across. And focally sat the Sun, in the same nearly central position that Ptolemy had gifted the Earth in his own special geocentric fudge, the last of the great Aristotelian models.

For the time being at least, Kant's 'island universe' theory had bitten the cosmic dust. In 1864, William Huggins had steered his spectroscope in the direction of the constellation Draco. He had wrongly concluded that the nebula found in that part of the sky was gaseous in nature, and hence analogous to our own Solar System. Indeed, by the late 1880s the 'island universe' theory had almost completely lost favour in the astronomical community. Though research programmes on the so-called spiral nebulae would eventually lead to the resurgence of the island universe theory for Einstein's sky, Wallace was more than able to discount it in his anthropocentric worldview.

So in some ways Wallace's anthropocentrism was ultramodern. The idea of the cosmos as a vast system of stars with our own Solar System at the centre was certainly *de rigueur* in many scientific circles. Wallace suggested that this galactocentric viewpoint had been arrived at after thirty years of data mining, and was now 'hardly questioned by any competent authority'.[9] The Solar System's central position in the star field was attained by painstaking research and observation. Only after meticulous star counts on both sides of the plane formed by the Milky Way found equality was the argument made for terrestrial centrality.

Our spherical 'solar cluster' was to be found within a general stellar system, which was also spherical. This cluster was defined by a group of a few hundred to a few thousand stars, surrounding our Sun. This central cluster seemed to condense in a group of stars separate from the rest of the Milky Way, which Wallace estimated to be around three thousand six hundred light years in diameter, as shown in Figure 16.

And so galactocentrism, and the centrality of the Sun, was the focal point of Wallace's cosmology. From this foundation he argued against the idea of life on other worlds. The Sun's centrality explained life on Earth too. From its central seat, the Sun was able to radiate a uniform glow over long periods of time for the creation and evolution of terrestrial life, which Wallace estimated at a few million years. The very power of the Sun remained a problem, of course. It was not until two further decades into Einstein's century that the atomic nature of solar radiation was appreciated.

Meanwhile, there was Kelvin. The Irish mathematical physicist William Thomson, otherwise known as 1st Baron Kelvin, was one of the main proponents of the meteoric hypothesis of the origin of the Sun's energy. A mere chemical approach had just not worked. When considered to be comprised of even the most

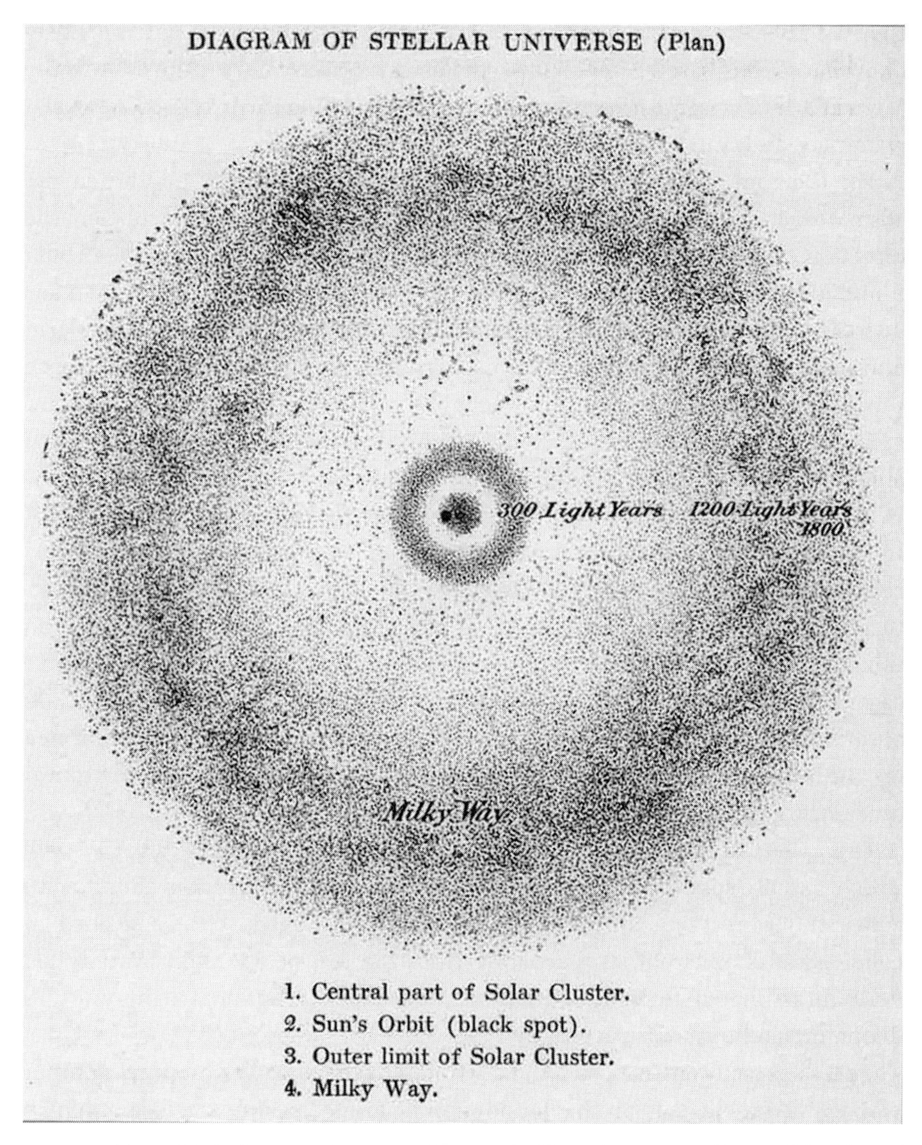

Figure 16 Anthropocentrism on a huge scale. The universe according to Alfred Russel Wallace, 1903, showing the Milky Way star system, 3600 light years across, and the Sun close to the galactic centre.

energetic of chemical reactions, the Sun would only supply around three thousand years' worth of heat. Under the meteoric theory, however, a comfortable twenty million years' supply was possible.

In Kelvin's view, the meteoric hypothesis was a simple case of the conservation of energy, albeit on a cosmic scale. When two or more meteoroids are separated in space they hold a significant amount of potential energy. And as they are drawn together by gravity, their energy is converted into kinetic energy and heat. Kelvin's solar theory simply deployed this meteoric hypothesis to include a larger number of meteoroids equal to the Sun's mass and mutually attracting each other to form the Sun. The same mechanism would stand for other stars in the same central position in the solar cluster, but more wayward stars would have shorter lives, unsuitable for the development of life on their attendant planets.

And so the greater number of stars that could potentially heat their planetary systems was reduced to a core few. In this galactocentric cosmology of Wallace, only a few hundred stars might have life-bearing planets in their orbits. And thus was contrived the first condition of Wallace's anthropocentric cosmology: that life can exist only around our Sun, or the cluster of suns in the solar neighbourhood.

Wallace's rare Earth hypothesis

One down, two to go. This left Wallace to explain the other two conditions of his anthropocentric cosmology: that no life exists on planets around other suns in the solar neighbourhood; and that no life exists in our Solar System beyond planet Earth. The second of these two conditions was familiar territory for Wallace as the associated arguments necessitated a discussion of biology. To understand life in space, one must first understand life on Earth, and as one commentator described Wallace, 'No great biologist has ever before seriously considered the possibilities of cosmic life, and they can only be fully discussed in the light of expert biological science'.[10]

And so Wallace brought his expertise to the question of pluralism. Though the field of biology was new, Wallace was daring. He began with a question that was to dominate the pluralist debate in Einstein's century, the question of life itself; what exactly *is* it? Rather than opting for the vagaries of philosophical discourse, Wallace homes in on the heart of the matter, the protoplasm. Described by Huxley as the 'physical basis of life', Wallace marvels at the complexity of the protoplasm, the living contents of all known organisms, which can nonetheless be broken down into the constituent elements of carbon, hydrogen, oxygen, and nitrogen.

Wallace's argument used a bottom-up approach. In his attempt to pinpoint the standard conditions for the development of microscopic protoplasm, Wallace drew rather cosmic conclusions. If the chemistry of the protoplasm determined the physical conditions needed for life on Earth, then from such small protoplasmic building blocks, far greater understanding could be achieved. Wallace listed these conditions as: a regular solar energy source, providing a restricted range of temperatures, with diurnal differences to keep temperature balanced; adequate heat and light; a prodigious and worldwide circulation of water; and a planetary atmosphere of requisite density.

Life, said Wallace, likes it cosy and snug. A temperature range between 0°C and 40°C is essential, along with sufficient solar energy to decompose carbon dioxide into carbon and oxygen for plant life. Wallace recognised that water was not only a major constituent of plant and animal life, but it also played a role in creating a limited range of temperatures. Meanwhile the planet's atmosphere must be necessarily dense to capture heat and provide oxygen, carbon dioxide, and water for life – a decrease of density by one quarter, says Wallace, would make the Earth unfit for life: 'Considering the great variety of physical conditions which are seen to be necessary for the development and preservation of life in its endless varieties, any prejudicial influences, however slight, might turn the scale, and prevent that harmonious and continuous evolution which we know must have occurred.'[11] And to reach such planetary conditions, the distance from the Sun, the obliquity of the ecliptic, the continuity of mild climate through deep time, and the circulation of water were all critical features.

Such detail had a cosmic purpose. Wallace's bottom-up approach highlights the complex conditions necessary to create and evolve life on Earth: but what of the planets beyond Earth? In stark contrast to the due consideration he gave to Earth's alleged special case, Wallace writes off planetary pluralism in a single sentence:

> The combination of causes which lead to this result [the development and maintenance of life over long periods] are so varied, and in several cases dependent upon such exceptional peculiarities of physical constitution, that it seems in the highest degree improbable that they can all be found again combined either in the solar system or even in the stellar universe.[12]

Wallace's arguments on man's place in the universe showed considerable cunning. He wanted to argue against life having evolved anywhere else in the Solar System, or indeed the cosmos. But Wallace was not content with the mere argument of the complexity and improbability of life ever evolving on Earth. He combined his biological reasoning with his cosmological case for special position, based on recent galactocentric observations.

Furthermore, recent advances in spectroscopy and the study of meteorites showed that the physics and chemistry of the cosmos was universal, 'We may therefore feel it to be an almost certain conclusion that – the elements being the same, the laws which act upon and combine, and modify those elements being the same – organised living beings wherever they may exist in this universe must be, fundamentally, and in essential nature, the same also.'[13] Though Wallace was aware of what he thought the consequences were, 'Within the universe we know, there is not the slightest reason to suppose organic life to be possible, except under the same general conditions and law which prevail here',[14] he did not speculate on whether extraterrestrials were composed of the same chemical elements as their terrestrial cousins.

Next, Wallace turned to the planets of the Sun. And here too he was on geocentric ground, arguing for a rare Earth. Wallace held that a planet's habitability was down to the mass of the planet. For this factor influences whether or not a planet can keep hold of the molecules necessary for an atmosphere. Mars and Mercury, then, would be too small to retain water vapour. And due to their low density, the larger planetary masses must have little solid matter on which life might begin and evolve. Only Venus remained. But as the satellite of love was thought to forever keep the same face toward the Sun, Venus was eliminated – one hemisphere too hot, the other too cold.

Nor did Wallace believe, in the lifetime of the Sun, that those planets currently unfit for life would ever have been fit in the past, or could ever be in the future:

> We are, therefore, again brought to the conclusion that there has been, and is, no time to spare; that the whole of the available past life-period of the sun has been utilised for life-development on the earth, and that the future will be not much more than may be needed for the completion of the grand drama of human history, and the development of the full possibilities of the mental and moral nature of man.[15]

Finally, Wallace purged the cosmos of candidate suns.

As he turned to the third condition, that no life exists on planets around other suns in the solar neighbourhood, Wallace engaged in a game of serial purging. Stars low in mass were purged from the useful list as they would fail to give enough heat. Stars in the Milky Way were purged due to 'excessive forces' in play in that region of the Galaxy. Thus was the cosmos reduced to those few stars of the solar cluster, 'variously estimated to consist of a few hundred or many thousand stars'.[16] But many of these too were purged. Some were considered too large or too small to gift the necessary enduring heat needed for the gestation of life. Others were in formation, and many others were trapped in binary systems, thought unfavourable for the development of planets. Thus the solar cluster, said Wallace, might hold but two or three single stars:

> But we do not really *know* that any such suns exist. If they exist we do not *know* that they possess planets. If any do possess planets these may not

> be at a proper distance, or be of a proper mass, to render life possible. If these primary conditions should be fulfilled, and if there should possibly be not only one or two, but a dozen or more that so far fulfil the first few conditions which are essential, what probability is there that all the other conditions, all the other nice adaptations, all the delicate balance of opposing forces that we have found to prevail upon the earth, and whose combination here is due to exceptional conditions which exist in the case of no other known planet – should all be again combined in some of the possible planets of these possibly existing suns? I submit that the probability is now all the other way.[17]

Wallace: the last great anthropocentrist?

And so Wallace orchestrated a symphony of anthropocentric refrains. But his arguments were based on the best science of the day. First, Wallace used galactocentrism to purge the cosmos of millions of candidate suns around which planets might evolve. Second, for those few suns left in Wallace's star atlas, he argued there was no life around these other stars of the solar cluster. Finally, he used his intimate and revolutionary knowledge of evolutionary biology. His conclusion, that the complexity of life on Earth made pluralism impossible, was made all the more potent for keeping company with his first two arguments. Wallace also attempted to show empirically that no planet in the Solar System other than Earth harboured life.

The philosophical angle too was tackled. In his pithy foray into the more theoretical side of the pluralist debate, Wallace proposed an idealist argument from purpose. It may be no surprise that the Earth is rare in the cosmos, suggested Wallace. The fact that our planet is the only one inhabited may be viewed as mere coincidence, or mightily significant. Wallace, of course, plumped for the latter. The universe, he proposed, was created for the supreme purpose of the evolution of man on Earth. Wallace was under no illusion that most scientifically minded people would call it coincidence. But for him, the opposite opinion was divergent from neither science nor faith. Life on numerous planets

> would introduce monotony into a universe whose grand character and teaching is endless diversity. It would imply that to produce the living soul in the marvellous and glorious body of man ... was an easy matter which could be brought about anywhere, in any world. It would imply that man is an animal and nothing more, is of no importance in the universe, needed no great preparations for his advent, only, perhaps, a second-rate demon, and a third or fourth-rate earth.[18]

Nor was the theme of time and space, soon so key in the unfolding narrative of Einstein's century, lost on Wallace. But the vastness of the emerging cosmos was viewed through his anthropocentric goggles. Space and time 'seem only the appropriate and harmonious surroundings, the necessary supply of material, the sufficiently spacious workshop for the production of that planet which was to produce, first, the organic world, and then, Man'.[19] Though such a teleological approach gets but a brief mention, it was the driving force in Wallace's philosophy for many years. As we discussed in the last chapter, Wallace created an unorthodox evolutionary theory, a kind of teleological evolution. His was an evolution of purpose and design, nutshelled in a theory that proposed all things were made for a final goal.

Wallace left his biological *coup de grâce* till last. This final evolutionary argument, which was quite compelling, was introduced into the 1904 edition of Wallace's book. It is particularly powerful as Wallace was recognised as one of the major figures in the development of evolutionary theory. And here, he applies it to the question of extraterrestrial life. Wallace argued that man is the product of a long series of modifications. Since these modifications occurred within a specific set of circumstances, and since the chance of those circumstances happening elsewhere in the cosmos are vanishingly small, the chance of beings like man on other worlds is risibly minute.

Furthermore, as no other creature on Earth had the intelligence of man, despite the huge number and variety of beasts that played out in evolutionary history, Wallace concluded that extraterrestrial intelligence was also vastly unlikely. Indeed, he was even prepared to number the improbability:

> If the physical or cosmical improbabilities as set forth in the body of this volume are somewhere about a million to one, then evolutionary improbabilities now urged cannot be considered to be less than perhaps a hundred millions to one; and the total chances against the evolution of man, or an equivalent moral and intellectual being, in any other planet, through the known laws of evolution, will be represented by a hundred million of millions to one.[20]

Wallace's contribution to the extraterrestrial debate was a lasting one. His definitive book on pluralism, *Man's Place in the Universe*, is idealist to the core. But the vulnerable teleological underbelly of his hypothesis was obfuscated by his use of cutting-edge astronomy, albeit galactocentrist, and evolutionary theory. The exacting and critical nature of Wallace's arguments on pluralism marks a significant advance in the debate, at a time when the science was about to dramatically change.

That change would come from Einstein. The sky in which Wallace drew his anthropocentric conclusions was still galactocentric. Einstein was about to re-introduce a decentralised cosmos. Won over by the near-central position of the Sun, Wallace plotted a dominant place for Earth and man in the cosmic scheme of things. But twentieth-century cosmology would reveal the true and terrible insignificance of Earth, in an unbelievably vast and new cosmos. Wallace contrasted the uniqueness of life on Earth with a barren universe, a realm whose sole existence was meant for the production of man alone. As Einstein's century wore on, such an idea became obscene.

Wallace was a curious case. Though his anti-pluralist rationale was based on materialist and scientific evidence of the Sun's centrality, the inventor of evolution was at heart an idealist. And as he took issue with the concept of the plurality of worlds, his belief in faith led to a justification of man's supremacy over nature. Once more, humanity was set apart. But as Einstein's century advanced, the material world would have its say.

Life in an expanding universe

On 26 April 1920, in the Baird auditorium of the Smithsonian Museum of Natural History, two of the world's most prominent astronomers were about to go into battle. Their prize: the scale of the universe. By all accounts, the debate on the day itself was something of an anti-climax. And yet the repercussions of its outcome, soaring way beyond the conference and into the scientific community, had far-reaching consequences for our understanding of space and time, and the development of life within it.

The contenders in the so-called Great Debate were US astronomers Harlow Shapley and Heber Curtis. Shapley was on his best behaviour. Present in the audience were delegates from Harvard who were about to interview him for the vacancy of Director of Harvard College Observatory, a post Shapley was to hold between 1921 and 1952. Shapley's view was that the Milky Way was the entire cosmos. His evidence: the work of the Dutch-American astronomer Adriaan van Maanen, whose conviction was that the rotation of galactic nebulae meant that their relative distances had been over estimated.

In opposition sat Heber Curtis. For Curtis the galactic nebulae were 'island universes', extragalactic civilisations of stars adrift in the black void of space. Working in the same tradition that had started with Kant, Curtis' position would soon be supported by the revolutionary work of fellow US astronomer, Edwin Hubble. Observations that Hubble made in the early 1920s soon led to the conclusion that the 'island universe' theory was correct. His findings would soon make headlines in the *New York Times* on 23 November 1924,

> Washington, Nov 22. Confirmation of the view that spiral nebulae, which appear in the heavens as whirling clouds, are in reality distant stellar systems, or 'island universes', has been obtained by Dr Edwin Hubbell [sic] of the Carnegie Institution's Mount Wilson Observatory, through investigations carried out with the observatory's powerful telescopes. The number of spiral nebulae, the observatory officials have reported to the Institution, is very great, amounting to hundreds of thousands ... The results are striking in their confirmation of the view that these spiral nebulae are distant stellar systems ... [and that] light ... has required a million years to reach us from these nebulae and that we are observing them from the Earth by light which left them in the Pliocene Age.[21]

But before Hubble, and before the Great Debate, there was Albert Einstein.

Two years after Wallace had penned his anthropocentric book on pluralism, Einstein had revolutionised physics in his 'miracle year' of 1905. He published four groundbreaking papers, the most notable of which was his theory of special relativity. A decade later, after years of intense effort and imagination, Einstein had another stunning success with his theory of general relativity, our modern description of gravity as a geometric property of space and time.

Spacetime was born. Einstein began to think about the implications of this latest theory for the universe as a whole. In 1917, he published a paper that launched a new field of physics, relativistic cosmology. Einstein wrote to his friend, the Austrian physicist Paul Ehrenfest, saying, 'I have ... again perpetrated something about gravitation theory which somewhat exposes me to the danger of being confined in a madhouse'.[22] Though Einstein had the daring to conjure up the first mathematical model of the cosmos, he was uncharacteristically cautious at a vital moment.

Rather than making one of the greatest scientific predictions in all of history, Einstein drew back from the brink. Instead of predicting Hubble's discovery of an expanding universe, a solution that naturally flowed from his field equations, Einstein elected to curb gravity and invent a new repulsive force. This force was quantified in his gravitational field equations by a parameter he called the 'cosmological term'.

Einstein was to describe it as the 'biggest blunder' he ever made. This confession was made to Russian physicist George Gamow and was symptomatic of the personal problems Einstein faced when he was writing his paper, 'Cosmological Considerations on the General Theory of Relativity'. But it was also symptomatic of the status of astronomy in 1917. At the time, as witnessed by the nature of the Great Debate, it was unclear whether anything lay beyond our Galaxy. The universe was still a very static place. So Einstein published a static solution for his relativistic cosmology.

But once more a realisation of the material nature of the universe, of matter in motion, would win through. By 1930 the observational work of astronomers like Hubble had made it clear the universe was expanding. Throughout the 1920s Hubble and other workers had used Cepheid variables, stars whose brightness varies according to a regular cycle, in Andromeda to deduce that the nebula was, at least, a huge one million light years away. And with his trusty colleagues, the wonderfully named Vesto Slipher and Milton Humason, Hubble observed 'red-shifted' recessional velocities on most galaxies, moving with speeds up to fifteen per cent of the velocity of light.

Einstein met with Hubble. In 1930 Einstein travelled to Mount Wilson, and became convinced of Hubble's evidence for an expanding universe, 'While on Mount Wilson, Einstein and his wife Elsa were given a tour of the observatory. It was explained to them that the giant telescope was used for determining the structure of the universe, to which Elsa replied: "Well, well! My husband does that on the back of an old envelope".'[23] And so Einstein abandoned his cosmological term: the galaxies were flying apart because the fabric of spacetime was spreading.

The part played by alien fiction in the space race

Einstein's curved spacetime gifted a *gedankenexperiment* to the writers of the world. His new sky became a theatre that unleashed the most incredible of concepts. A new stage was set, which included not just all of space, but all of time too. The astounding age of science fiction, those first few decades of the twentieth century, featured the pulp fiction of dizzying technologies, immense interstellar starships, and fleet and nimble faster-than-light vessels worming through spacetime. The stories that at first were merely pulp would later prove seminally influential on the space race.

Such was the synergy of science fiction and science.

In this age of Einstein, with a new-found frontier stripped of spatial and temporal parochialism, anything seemed possible. And as the physics fed back into fiction, some of the classic authors of the late nineteenth century were to have a profound effect on the twentieth. Consider American rocket pioneer Robert Goddard. The motive force for Goddard was not the arcane glyphs and graphs of science, but the fiction of H. G. Wells. As Goddard confessed in a letter to Wells,

> In 1898, I read your *War of the Worlds*. I was sixteen years old, and the new viewpoints of scientific applications, as well as the compelling realism

> . . . made a deep impression. The spell was complete about a year afterward, and I decided that what might conservatively be called 'high altitude research' was the most fascinating problem in existence.[24]

Inspiring fiction was also the galvanising factor for German rocket pioneer Herman Oberth. At the age of eleven, Oberth's imagination was sparked by Jules Verne's *From The Earth To The Moon*, a book that he later recalled he had read 'at least five or six times and, finally, knew by heart'.[25] And among those inspired in turn by Oberth was his star pupil, Wernher von Braun. A controversial figure by most measures, von Braun was the lead engineer behind the Nazi 'retaliation weapon', the V2 rocket. Von Braun became naturalised in the United States following World War II as part of their space programme. This despite controversy over his status as an SS officer and his use of concentration camp prisoners as labour, which saw more people killed by the production and delivery of the V2 than by their actual deployment. Almost twenty-five years after abandoning Germany, the von Braun-designed Apollo 11 took off for the Moon, propelled by the Saturn V booster. Rarely in the record of human enterprise has so stirring an aspiration been achieved by such ethically dubious means.

Thus these early rocket pioneers, so influenced by early alien fiction, were instrumental in developing the means for the actual exploration of space. Indeed, the rocket pioneers fed their expertise back into fiction. Austro-Hungarian film-maker Fritz Lang used von Braun and Oberth as technical advisors for his groundbreaking science fiction film *Frau Im Mond* (*Woman in the Moon*). During filming, in an attempt to build a working rocket to drive publicity for the film, Oberth lost the sight in his left eye. The injury was typical. A long list of such casualties litters the history of early rocketry, and so we envisage a motley collection of revenants hoping to keep both body and mind together whilst lurching towards their final goal.

In the East too the influence of fiction was profound. Take the father of Russian rocketry, Konstantin Tsiolkovsky. Tsiolkovsky's most famous work was *The Exploration of Cosmic Space by Means of Reaction Devices* (1903), which is the first scholarly treatise on rocketry. And yet an integral part of his entire approach to the question of rocketry and space travel was the power of fiction. During his lifetime, Tsiolkovsky authored over five hundred works on space travel and related subjects. Among them were his science fiction novels, including *On The Moon* (1895), *Dreams of the Earth and Sky* (1895), and *Beyond the Earth* (1920).

Tsiolkovsky's work may seem at first to meander across the whole field of rocketry without focusing upon any one area. Nestling amidst the designs of rockets are plans for steering thrusters, multi-stage boosters, space stations, airlocks, and essential systems to provide for the successful operation of space colonies. The whole body of work was united by Tsiolkovsky's mission: that inexpensive and regular access to space and the other planets would make it possible for humanity to extend its reach into space.

Tsiolkovsky began in pulp. His first fiction was a short story, *On the Moon*, published in a Moscow magazine in 1892.[26] *On the Moon* was a taste of things to come: an attempt to make manifest the transhuman aspirations that were also found in Tsiolkovsky's science writing. The alien is explored in his stories in a similar way to other tales throughout the world of mass magazines of the late nineteenth and early twentieth centuries. Together these new periodicals, especially the pulp fiction magazines of science and science fiction, published many stories of scientific romance and wild adventure, such as Edgar Rice Burroughs' famous *Mars* series.

And so Tsiolkovsky's charge was to plan a cosmic human future, and this calling always lay at the core of his work, fact and fiction. It was an approach he outlined in detail in his 1926 *Plan of Space Exploration*, consisting of sixteen steps from the very beginning of space conquest, until the far distant future, including interstellar travel:

- Creation of rocket airplanes with wings
- Increase of the speed and altitude of these airplanes
- Production of real rockets, without wings
- Ability to land on the surface of the sea
- Reaching escape velocity . . . and the first flight into Earth orbit
- Lengthening rocket flight times in space
- Experimental use of plants to make an artificial atmosphere in spaceships
- Using pressurised space suits for activity outside of spaceships
- Making orbiting greenhouses for plants
- Constructing large orbital habitats around the Earth
- Using solar radiation to grow food, heat quarters, and for transport through the Solar System
- Colonisation of the asteroid belt
- Colonisation of the entire Solar System and beyond
- Achievement of individual and social perfection
- Overcrowding of the Solar System and the colonisation of the Milky Way Galaxy
- The Sun begins to die and the people remaining in the Solar System's population go to other suns

Last and First Men

The evolution of mankind into a starfaring species outlined by Tsiolkovsky was a concept explored in much greater detail in the work of Olaf Stapledon. Born in 1886 in the Liverpool area of England, Stapledon was a philosopher and an innovative and influential science fiction author. Awarded a PhD in philosophy from the University of Liverpool in 1925, he soon turned to fiction in the hope of presenting his ideas to a wider public.

For astrobiology, Stapledon produced two key works. The first was *Last and First Men* (1930), an anticipatory history of eighteen successive species of humanity. The second, *Star Maker* (1937), was an outline history of the universe. Stapledon's writings greatly influenced not only key players in our own story on pluralism, such as Arthur C. Clarke and Stanislaw Lem, but also figures as diverse as Jorge Luis Borges, Bertrand Russell, Virginia Woolf, and Winston Churchill.

Stapledon was a writer equipped for the age of Einstein. These two main volumes from Stapledon's collection of books on fiction and philosophy synthesised astronomy and evolutionary biology. In turn they conjured up cosmic myths suitable to a sceptical and scientifically cultured age. As American author and astrophysicist Gregory Benford comments in his introduction to *Last and First Men*:

> Stapledon had studied the principal scientific discovery of the nineteenth century, the Darwin–Wallace idea of evolution and projected it onto the vast scale of our future, envisioning the progress of intelligence as another element in the natural scheme. In his hands pressures of the environment become the blunt facts of planetary evolution, the dynamics of worlds.[27]

Evolution and Einstein had transformed the concept of the alien. And as biology and relativistic cosmology found their way into twentieth-century fiction, the alien became one of the enduring and universal motifs of the age. Consequently, an increasing number of people met these revolutionary ideas not through science, but as a text, inspiring emotional as well as intellectual reactions. Thus, ideas of life in the new universe were embedded more deeply into the public psyche, and, to echo a quote we met in Chapter 1, scientific ideas were creatively morphed into symbols of the human condition that were often

> an unconscious and therefore particularly valuable reflection of the assumptions and attitudes held by society. By virtue of its ability to project and dramatise, science fiction has been a particularly effective, and perhaps for many readers the only, means for generating concern and thought about the social, philosophical and moral consequences of scientific progress.[28]

In turn, as scientists are creatures of the culture in which they swim, alien contact narratives of the late nineteenth and early twentieth centuries motivated a significant number of scientists. The idea of life in the universe, and man's place within it, was firmly fixed in the scientific as well as the popular imagination. The theory of evolution had given credence to the evolution of life on Earth, and to evolution in a cosmic setting. The new sciences of biology and cosmology inspired a wealth of fiction,[29] and provided a rationale for imagining what cosmic life might develop. From now on the idea of cosmic life became synonymous with the physical and mental characteristics of the alien. It provided a new rubric against which man himself could be measured.

As we have seen, the impact of H. G. Wells was colossal. *The Time Machine* (1895) and *The War of the Worlds* (1898) were responsible for igniting both the space and time themes in the genre of science fiction, and in the public imagination. Wells created the modern nexus of the alien, armed with its potential for probing human evolution. And Wells' books 'are, in their degree, myths; and Mr Wells is a myth-maker'.[30]

Once developed by Wells, the modern alien idea proved a potent motif for cultivating fictional explorations of the singularity or insignificance of humanity. During such explorations, the secondary question of the character of alien and interspecies interaction became an issue, which later affected the Search for Extraterrestrial Intelligence (SETI) science programme. As English science fiction writer Brian Aldiss put it, 'Wells is the Prospero of all the brave new worlds of the mind, and the Shakespeare of science fiction'.[31]

If H. G. Wells is the Shakespeare of science fiction, then Olaf Stapledon is its Milton. Stapledon used the science fiction genre to explore nothing less than the meaning of human existence in a cosmic setting. And his books attempt to imagine and project the prospect of physical contact with alien life. Witness Stapledon's account of the Martian invasion of Earth. It is typical in the way that Stapledon's humans and aliens alike cross the void between the stars in vessels unimaginable to moribund Earth dwellers. Viewed from the confines of a contemporary pedestrian existence, Stapledon's vessels journeyed the skies above in both architecture inconceivable and technology incredible. Science fiction had invented the UFO:

> Early walkers noticed that the sky had an unaccountably greenish tinge, and that the climbing Sun, though free from cloud, was wan. Observers were presently surprised to find the green concentrate itself into a thousand tiny cloudlets, with clear blue between ... though there was much that was cloudlike in their form and motion, there was also something definite about them, both in their features and behaviour, which suggested life.[32]

In the preface to *Last and First Men*, Stapledon tells the reader that his story is an attempt 'to see the human race in its cosmic setting, and to mould our hearts to entertain new values'.[33] In a telling evocation of the theories of evolution and relativity, he suggests that such attempts to extrapolate man's evolutionary future 'must take into account whatever contemporary science has to say about man's own nature and his physical environment'.[34] Stapledon produced a fiction that incorporated the most recent ideas of cosmology and evolutionary biology. Thus he created a new fusion of fact and fiction, a form of fable for a scientifically cultured twentieth century. In the words of Stapledon, the aim must not be just 'to create aesthetically admirable fiction ... but myth'.[35]

That Stapledon was well aware of the latest developments in astronomy is evident from his one of his visions on the human future in space:

> But in the fullness of time there would come a far more serious crisis. The sun would continue to cool, and at last man would no longer be able to live by means of solar radiation. It would become necessary to annihilate matter to supply the deficiency. The other planets might be used for this purpose, and possibly the sun itself. Or, given the sustenance for so long a voyage, man might boldly project his planet into the neighbourhood of some younger star ... He might explore and colonize all suitable worlds in every corner of the galaxy, and organize himself a vast community of minded-worlds.[36]

Stapledon's imagined life in the universe is one of emerging genetics. He imagined the future forms of man, in an alien setting. *Last and First Men* is a future history on a staggering scale. The 'hero' of the book is not a man, but mankind. The story embraces seventeen evolutionary mutations, from the present 'fitfully-conscious'[37] First Men, to the glorious godlike Eighteenth Men who reign on Neptune. It is a history that spans two thousand million years.

Eugenics was the dark side of Darwinism. The first and most infamous exponent of a genetic intervention in the human race was Darwin's cousin, Francis Galton. Indeed, it was Galton who had introduced the word eugenics into the lexicon. In *Last and First Men*, Stapledon thought long and hard about eugenic practices. One of the causes of the demise of the First Men, for instance, was their failure to realise a eugenics programme:

> In primitive times the intelligence and sanity of the race had been preserved by the inability of its unwholesome members to survive. When humanitarianism came into vogue, and the unsound were tended at public expense, this natural selection ceased. And since these unfortunates were incapable alike of prudence and of social

> responsibility, they procreated without restraint, and threatened to infect the whole species with their rottenness.[38]

So, human intelligence steadily declined, 'And no one regretted it'.[39]

Later came the irresistible rise of the Third Men. With their rediscovery of eugenics, the Third Men focused their efforts on that most distinctive feature of

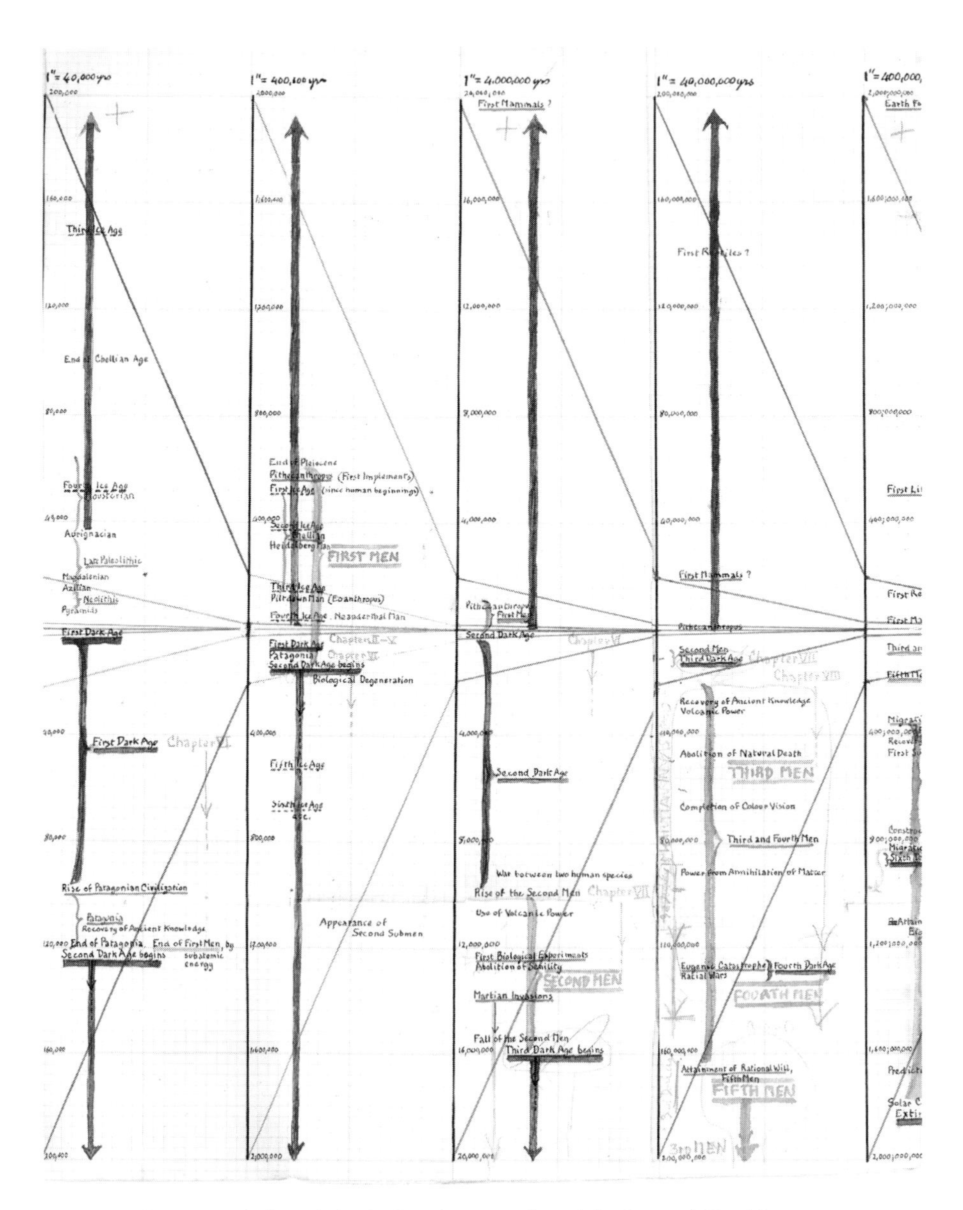

Figure 17 Olaf Stapledon's timechart storyboard for *Last and First Men.*

man, the mind. Seeking to 'breed strictly for brain, for intelligent coordination of behaviour',[40] the peak of their engineering programme was the Great Brains.

The Great Brains first helped, then enslaved, and finally annihilated their creators. Ultimately they turned their cool intellects upon themselves. They created a superior species, the Fifth Men. The Fifth Men were accomplished in art, science, and philosophy, perfectly proportioned of body and mind. They were able to travel mentally back through time to experience the whole of human existence. Indeed, the Fifth Men became the most perfect species ever to dwell on planet Earth.

Star Maker

The conceptual reach of *Star Maker* beggars belief still further. Its scope is so vast that *Last and First Men* would warrant hardly a mention in the cosmic sweep of Stapledon's next book. As one commentator put it, 'The truly incredible scope of *Star Maker* becomes apparent only in the latter half of the book. Here, the entire two billion year history of man that is recounted in the three hundred or more pages of *Last and First Men* is reduced to a single paragraph.'[41] The purpose and presence of the alien in *Star Maker* is, in the words of Stapledon himself, to 'explore the depths of the physical universe [and] discover what part life and mind were actually playing among the stars'.[42]

So Stapledon broadens his vision, and takes into his telescopic field not only the entire history of the universe, but the history of other universes too. To the reader of his day, the possible existence of sequential universes must have been mind-boggling. His 'Star Maker' is not God, in the common sense, but a supreme artist, a creator of successive universes until the ultimate universe perfectly conveys the spirit of its creator. Our entire universe, as it evolves, is nothing but a sketch, a rough draft, for 'the final of the ultimate cosmos'.[43]

In some ways *Star Maker* started where *Last and First Men* ended. An anonymous narrator, the ethereal 'I' of the story, falls into a trance-like state while meditating on a hill next to his home. The quintessence of 'I' is swept off Earth, through the Solar System and beyond, ever faster. Kepler's protagonist in *Somnium* had been spirited away to the Moon by demons. Stapledon's is a new kind of journey of the spirits.

It is a voyage through Einstein's sky. The new generation of Einstein had brought knowledge of a new universe. And as Einstein's equations were made flesh by deep space astronomers, Stapledon was ready with a new model, and a new myth, for Einstein's universe. The ethereal 'I' travels on, in quest of planets that might harbour humanoid life. Eventually our narrator sets down on Other Earth, a planet described in some detail, in terms of society and structure.

'I' moves on, using faster-than-light travel, and reports a poetic vision of the Doppler effect, 'since I was travelling almost in the plane of the Galaxy, the hoop of the Milky Way, white on either hand, was violet ahead of me, red behind'.[44] Myriad worlds are explored. Myriad modes of life. Stapledon conjures up an incredible variety of species of men, far more even than in *Last and First Men*.

A procession of 'strange mankinds' is paraded before the reader. Portraying an array of different alien conditions, Stapledon conjures up a teeming variety of cosmic life: centaurs and echinoderms, intelligent spaceships and multiple minds, composite beings and symbiotic races. The journey takes us through terrible crises in galactic history, planetary wars, galactic empires, faster-than-light travel, and even telepathic sub-galaxies descending into insanity.

Stapledon also explores the pantheistic possibilities of the cosmos. In a spiritual examination that began with Giordano Bruno's idea that planets and stars have souls, Stapledon makes the heavenly bodies mindful. Stars have mentalities and their minded planets are able to communicate with them. As the galaxy falls into decay, there is at least a symbiotic relationship between stars and planets. As the narrator reports, 'the great snowstorm of many million galaxies',[45] a full telepathic exploration of the universe becomes a possibility. And yet the 'I' is still a mystery to itself.

The sheer scale of the narrative is ramped up once more as the 'I' becomes part of the cosmic mind. The 'I' listens to the inner thoughts of the nebulae as it continues to search for the Star Maker itself. And as this Supreme Creator is finally found, we discover it to be stellar, and remote. The Star Maker is repulsed by the raptures of the cosmic mind. As British science fiction writer, Brian Aldiss, comments on the Star Maker, 'the created may love the creator but not vice versa, since that would merely be self-love of a kind'.[46]

A dream is induced in the cosmic mind. Before this cosmos, there were others. The young Star Maker had experimented on early models of the universe, now tossed into the dustbin of cosmic history. Of these early versions, one such cosmos featured three universes, linked as one. In the first of these three universes, a couple of spirits, one good and one bad, play for the possession of the souls of creatures. Whether they win or lose determines the destiny of the creatures' souls, which are plunged either into an eternal heaven, or an eternal hell, in one of the other universes. In our cosmos, we retain a distant memory of these ancient domains.

Like an engineer tinkering with the software of his cosmos, the Star Maker continues to refine versions of his design, as his own skills develop. With tragic persistence, the Star Maker is forever outgrowing his creations. Though later versions show better design, flaws and suffering are always rife. At last, the cosmic mind is exhausted by the procession of realms, and awakes from the 'dream', realising in fact that it had been contemplating the consciousness of the Star Maker.

And here is the meaning of life in the universe. The Star Maker is able to know all realms and all life in one timeless vision. Such knowledge is the greatest goal of the universe. With this understanding, the 'I' returns to Earth, where he takes comfort in the parochial limits of the present. But not for long, as war looms in Europe, and there is little time for respite.

These two books of Stapledon were anathema to the devout. *Last and First Men* reads like an atheist manifesto. And in *Star Maker*, as Aldiss points out, 'the atheism has become a faith in itself, so that it inevitably approaches higher religion, which is bodied forth on a genuinely new twentieth-century perception of cosmology'.[47] In Aldiss' view, *Last and First Men* and *Star Maker* turn scientific ideas into vast and epic prose poems, and Stapledon is the ultimate science fiction author.

Star Maker is the first classic odyssey of the twentieth century.

It recalls another famous and classic journey, the travels of Dante Alighieri through the worlds of Hell, Purgatory, and Paradise in the three books of his *Divine Comedy*. But whereas Dante journeys through the medieval Christian universe, adopted from Aristotle, Stapledon journeys through Einstein's sky. Witness the darkness and confusion at the beginning of the book, reminiscent of Dante's *Inferno* where Dante is lost in a dark wood, but Stapledon tells his own tale:

> One night when I had tasted bitterness I went out on to the hill. Dark heather checked my feet. Below marched the suburban street lamps. Windows, their curtains drawn, were shut eyes, inwardly watching the lives of dreams. Beyond the sea's level darkness a lighthouse pulsed. Overhead, obscurity.[48]

As with Dante, Stapledon's book is an exploration of life in the universe. In both books, the narrator moves from the dark world of his soul to an all-embracing revelation of the cosmos and its creator. Stapledon's narrator is guided on his journey by an advanced extraterrestrial intelligence on a distant world. And the various worlds he visits on the way bear similarity to the afterworlds visited by Dante.

But there the journeys diverge. The final revelation of *Star Maker* is radically different from Dante's *Divine Comedy*, 'as Dante's vision was essentially comic because God cared for man, so Stapledon's is ultimately tragic, because the Star Maker cares only for creation and the critical contemplation of his creatures'.[49] *Star Maker* is, from another point of view, a pluralist account of Einstein's pantheist universe. It is a Divine Tragedy. While Dante's epic poem is rooted in the Christian idea of good and evil, Stapledon's account of life in the universe is a more morally ambiguous one. We are confronted with a more complex case in which we learn that the force behind the cosmos, the Star Maker, transcends the narrow bounds of good and evil. And 'I', the narrator, made captive by the confines of his own perspective, tries in vain to understand the Star Maker as an embodiment of good or evil. As he is joined by other cosmic explorers in this quest, however, 'I' learns that Star Maker is something quite alien.

Stapledon's literary work is also the latest instalment in science fiction's forward thinking. The setting in which *Star Maker* was conceived had undergone a further, though more silent, revolution in cosmology. It was not until the late 1950s that astronomers drew analogies between revolutions in cosmology and the impact of finding extraterrestrial intelligence.[50] Woefully slow when compared to fiction. After all, Kepler's writing of *Somnium* in 1634 directly implied such a conclusion! Nonetheless, professional astronomers suggested that alien contact would represent 'The Fourth Adjustment' in humanity's outlook, following the shift to the geocentric, heliocentric, and galactocentric worldviews.[51] And as we have seen, this latter revolution, hastened by discoveries showing that our Solar System was merely at the edge of our Galaxy, and that the Galaxy itself was but one of many, was emerging just prior to the time Stapledon was writing *Star Maker*.

By the days of Einstein's sky, astronomy had undergone great revolutions:[52] the Copernican, the galactocentric, and Einstein's prediction of Hubble's expanding universe. But there was one massive upheaval yet to come; the answer to the question 'Are We Alone in the Universe?' The revolution had already begun with Kepler, and continued with Stapledon. By the late 1920s, revolutions, including those of Copernicus, Darwin, and Einstein, may have inured the masses to marginalisation.[53] But Stapledon was preparing the public for the final great demotion, and in the process helped develop the myth of the close encounter of the third kind: physical contact.

In *Star Maker*, alien biologies, together with terrestrials, search for the supreme intelligence in the new universe. Stapledon's story is a search for the holy grail of Einstein's sky. It is a quest for the spirit of the cosmos, an entity at the head of a new, and cosmic, Great Chain of Being. In an early evocation of the implicit inhumanity of the new universe, Stapledon writes, 'it was becoming clear to us that if the cosmos had any lord at all, he was not that spirit [God], but some other, whose purpose in creating the endless fountain of worlds was not fatherly toward the beings that he made, but alien, inhuman, dark'.[54]

Stapledon's fiction, then, emphasised the triviality of humanity in the face of a new and vast cosmos, which itself may harbour truths and meaning as yet unknown to an immature human race. His fiction on the question of intellectual contact with alien biologies had a lasting influence on the twentieth century. Working scientists, such as J. B. S. Haldane, physicist Fred Hoyle, and Carl Sagan, one of the founders of SETI in the early 1960s, were swayed by Stapledon, as was fellow British science fiction writer, Arthur C. Clarke.

The origin and nature of life

One of the major influences on the fiction of Olaf Stapledon had been British-born geneticist and evolutionary biologist, J. B. S. Haldane. Of aristocratic family, and educated at Eton and Oxford, Haldane dedicated his life to the study of nature. He was a staunch materialist, and devoted to empirical evidence, believing that atheism was the only rational deduction: 'My practice as a scientist is atheistic. That is to say, when I set up an experiment I assume no god, angel or devil is going to interfere with its course ... I should therefore be intellectually dishonest if I were not also atheistic in the affairs of the world.'[55]

Haldane was prolific in both fact and fiction on the question of life in the universe. Most notable was his essay on the origin of life. Haldane's account had come independently of another worker in the field, Russian biochemist Aleksandr Oparin. Together their thoughts became known as the Oparin–Haldane theory of the origins of life. The vision was one of an early Earth, with a reducing atmosphere and a primitive ocean, teaming with organics that had, in Haldane's words, 'reached the consistency of hot dilute soup'.[56]

But a greater influence on Stapledon than the primordial soup theory had been Haldane's essay about the end of the world. *The Last Judgement*, published in 1927, was an astonishing essay from a prominent scientific worker at the cutting edge of the biochemistry of life. Haldane conveyed a vision of a human future over the next forty million years. He updated H. G. Wells' futurism with the long-range repercussions of the 'new biology' on human fate. It was a vision that was to inform Haldane's scientific work for the rest of his life.

Throughout the first half of the twentieth century, the question of the nature and origin of life, terrestrial and extraterrestrial, remained a key issue for the pluralist debate. One of the main contributors to the debate was prominent Irish physicist, J. D. Bernal. His *Physical Basis For Life* (1951) attempted to address the question of 'what is life?', a question even more crucial when applied to an extraterrestrial context. Bernal commented in his 1947 lecture that, unlike physics and chemistry,

> terrestrial limitations obviously beg the question of whether there is any more generalized activity that we can call life ... Whether there are some general characteristics which would apply not only to life on this planet with its very special set of physical conditions, but to life of any kind, is an interesting, but so far purely theoretical question.[57]

In a 1952 lecture to the British Interplanetary Society, Bernal extended his remit to include the origin of life in the universe, contending, 'the biology of the future would not be confined to our own planet, but would take on the character of cosmobiology'.[58] Bernal remembered that when he conferred with Einstein on the question, the latter had suggested that any universal description of life would be more a matter of poetry than fact: 'Any self-subsisting and dynamically stable entity transforming energy from any source, or as Haldane put it, "any self-perpetuating pattern of chemical reactions," might be called alive in this sense'.[59]

An evolutionary approach to the question of the form and nature of alien life, and alien intelligence, had been broached in fiction since the end of the previous century. French writers in particular had imagined a cosmobiology, which fitted humans and aliens alike into a great evolutionary scheme. The work of Camille Flammarion is a case in point, wherein are descriptions of sentient plants, species for which respiration and alimentation are facets of the same process. Flammarion's *Urania* (1889) had promoted the idea that souls could undergo serial reincarnation in an infinite variety of physical forms.

Aliens of unusual form also appeared in the works of J. H. Rosny aîné, a French writer of Belgian origin. Rosny was the pseudonym of Joseph Henri Honoré Boex, an author considered in that part of the world as one of the founding figures of modern science fiction, and perhaps second only to Verne in the eyes of the French. Rosny's work did not lack imagination. Inorganic aliens in the form of minerals featured in *The Shapes* (1887) and *The Death of the World* (1910). And in *The Navigators of Infinity* (1925), way before the days of Kirk on Star Trek, there is a love affair between a human and a six-eyed tripedal Martian.

The Black Cloud

Fictional aliens began to take on more imaginative forms.

Fred Hoyle was a British physicist of considerable reputation and renown in the post-war scientific community. In time, Hoyle was to occupy the very illustrious Plumian Professorship of Astronomy and Experimental Philosophy at the University of Cambridge, and later still become the founding director of the Institute of Theoretical Astronomy, where under Hoyle's leadership the institution would quickly become one of the leading groups in the world for theoretical astrophysics.

Winding back the clock to the 1950s, Hoyle had been an early champion of plentiful solar systems, and the likelihood of alien life. He had authored potentially Nobel Prize-winning work on the theory of stellar nucleosynthesis (his group being the first to realise the fundamentals of how the chemical elements of the cosmos are first fused). And it was Hoyle who coined the name Big Bang, when referring disparagingly to the standard cosmology as this 'big bang idea' in a radio broadcast on 28 March 1949, on the BBC Third Programme. He went on to appear in a series of legendary talks on astronomy for the BBC, which were later collected in a book, *The Nature of the Universe*, the first in Hoyle's series of popular science books.

Indeed, Hoyle was the architect of an entire cosmology. His 'steady state' view of the cosmos was an alternative to the developing standard cosmological model of the Big Bang. Hoyle had no argument with an expanding cosmos. What he objected to was the interpretation that the universe had a beginning. It was, like Judeo-Christian religion, linear in time. For Hoyle, a life-long atheist, anti-theist and Darwinist, it was pseudoscience. The ignominy of a cosmology embraced by the Church inspired Hoyle to search for an alternative. Like a river whose molecules were in motion, but whose overall body flowed as one, the universe described by the steady state cosmology was eternal and immutable, while its constituent galaxies were in constant motion.

So Hoyle was very much at the bleeding edge of cosmology, as well as pluralist fact and fiction. In his first novel, *The Black Cloud* (1957), Hoyle explored similar ideas about disembodied intelligence to those of Stapledon before him. The alien intelligence in Hoyle's book takes on the form of a cloud of interstellar matter, some five hundred million years old, with which astronomers are ultimately able to communicate.

In the preface to *The Black Cloud*, Hoyle makes the point that 'there is very little here that could not conceivably happen'.[60] He speaks not only of his cosmology behind the fiction, but also the contemporary context in which the book was written. Set barely a decade into Hoyle's future, 1965–75, the book's narrative purports to be an account discovered, many years after the events described, among the private papers of one of the major players in the story. The world in which the story is set is one familiar to Hoyle: an Oxbridge circle of eminent scientists, and the 'two cultures' soon to be described so famously by C. P. Snow.

Besides pluralism, *The Black Cloud* also explores politics. Through its main protagonist Kingsley, a man largely like Hoyle himself, the book explores the social responsibility of science and the state. Kingsley is a maverick but brilliant astrophysicist, a Cambridge professor, who is regarded by the politicians who fund his research work as something of a rebel. Kingsley has little patience with the apparatchiks, and prefers, like Wells before him, the idea of a scientific meritocracy, where the future of man is left in the hands of the international brotherhood of science. Women barely register on the scientific scale of the story.

The story begins with the discovery of the Cloud, and its implications for science. Not only does Hoyle describe the scientific theorising about the nature and speed of approach of the Cloud, he also focuses on the political suspense that unfolds as world governments come to terms with the consequences of the Cloud's imminence. Hoyle's main exploration of the science of the alien is the question of the form and nature of life. And in that he focuses on the physical and psychological nature of the Black Cloud itself. In the book's preface, Hoyle refers to the Cloud as a 'black hole in the sky', a subject still regarded as speculative at the time and not developed until Roger Penrose's seminal paper in 1965 on 'Gravitational Collapse and Space-Time Singularities'.

Into this emergent concept of a black hole, Hoyle builds the idea of the Cloud as a well of alien intelligence. It is the ultimate source of transmissions, which at first are thought to come from the ionic fluctuations at the periphery of the Cloud. As the energy to produce such ionisation is otherwise insufficient, the conclusion must be that the Cloud is the source of power, able to react and communicate.

In *The Black Cloud*, Hoyle also takes the ideas of the likes of Haldane into fiction. Hoyle's narrative speculates on the difference between the animate and inanimate, the organic and the inorganic. However, Hoyle's fiction is so meticulous in its authenticity that he uses footnotes to give calculations for specific problems in the text. The classification of the Cloud itself is an example of the way in which the fact/fiction interface is characterised.

As the Cloud approaches, the scientific reaction is portrayed as dialectic and changing in nature. A range of hypotheses are presented on the question of the Cloud, and they range from anxiety over possible atmospheric heating and cooling, through the biological properties of plants and animals, to the Newtonian laws of celestial mechanics. Ultimately, the Cloud arrives in the Solar System. It lurks just in the right location to obscure the Sun's light. And as the Earth suffers colossal ecological repercussions caused by the radical climactic changes, the scientists are bunkered away, one step removed from the drama that unfolds.

And yet the Cloud makes its mark, even on the isolated boffins. The psycho-political effect on the scientists first shows when they become more self-reliant, much to the chagrin of the politicians. Next, the scientists wrestle power, as they possess the only radio equipment on Earth capable of universal transmission and reception. At last, they discover the true nature of the Black Cloud. It is an incalculably ancient being, and vast in intellect. Their knowledge and hardware enables them to make contact with the Cloud, and communication begins.

The scientists confer with the Cloud, and ask questions on life in the universe. They query the nature of God, and the origin of life itself. Such questions trigger in the Cloud an admission that it is presently leaving precisely for the same questions. News of an event occurring in the solar neighbourhood will hopefully throw light upon answers to such questions. But before it investigates, it prepares to communicate to the scientists fundamental knowledge of the cosmos, but *in its own language*.

In his account of the dialogue with the Cloud, Kingsley assumes himself to be free of geocentrism, free of anthropocentrism, and free of the psychological hurdles associated with Earth-centredness. It becomes clear, however, that geocentrism still pervades his thinking, as Kingsley's science is still wrapped up in the anthropocentric laws applicable only to Earth. And so the context of Kingsley's science is set. His account of the genesis, evolution, and physiology of the Cloud is fundamentally flawed by its being based on earthly beasts: evolution takes place within the Cloud; its genesis explained within the confines of terrestrial biology; its physiology in terms of magnetism; and its neural pathways in terms of terrestrial laws of electromagnetism.

In the end, geocentrism is the killer. Anthropocentrism destroys Kingsley, as the esoteric nature of the data transmitted into his brain by the Cloud was so alien it radically altered his neural networks, swiftly ensuring his death. With the Cloud gone, and Kingsley a dead scapegoat, life, politics, and science return to normal. But *The Black Cloud* is the shape of things to come. As the age of SETI dawned, writers such as Stanislaw Lem and Arthur C. Clarke were to take the thoughts of Stapledon and Hoyle further still. And Hoyle was to feature in an earthly conflict of the cosmos, as the battle between the steady state and the Big Bang saw cosmology become a science.

Notes

1 Stapledon, O. (1930) *Last and First Men*, London, Gollancz edition, SF Masterworks (1999) Chapter 15

2 Wallace, A. R. (1903) Man's Place in the Universe, as Indicated by the New Astronomy, *The Fortnightly Review*, 1 March, 1903, 396

3 Stapledon, O. (1937) *Star Maker*, London, Gollancz edition, SF Masterworks (1999) Chapter 6

4 Amis, M. (1995) *The Information*, London, p. 93

5 Sagan, C. (1994) *Pale Blue Dot*, New York

6 Huxley, T. H. (1863) *Man's Place in Nature*, Ann Arbor, Mich., p. 71

7 Dick, S. J. (1996) *Biological Universe*, Cambridge, p. 15

8 Wallace, A.R. (1903) *Man's Place in the Universe: A Study of the Results of Scientific Research in Relation to the Unity or Plurality of Worlds*, New York, Preface, pp. v–vi

9 ibid., p. 165

10 Quoted in Dick, S. J. (1996) *Biological Universe*, Cambridge, p. 44

11 Wallace, A. R. (1903) *Man's Place in the Universe: A Study of the Results of Scientific Research in Relation to the Unity or Plurality of Worlds*, New York, pp. 191–4
12 ibid., p. 310
13 ibid., pp.182–9
14 ibid.
15 ibid., p. 275
16 ibid.
17 ibid., pp. 278–85
18 ibid. pp. 317–18
19 ibid.
20 Wallace, A. R. (1904) *Man's Place in the Universe*, London, Appendix, pp. 334–5
21 Sharov, A. S. and Novikov, I. D. (1989) *Edwin Hubble: The Discoverer of the Big Bang Universe*, Cambridge, p. 34
22 Quoted in Hey, T. and Walters, P. (1997) *Einstein's Mirror*, Cambridge, p. 210
23 Berendzen, R., Hart, R., and Seeley, D. (1976) *Man Discovers the Galaxies*, New York, p. 200
24 Goddard, E. C. and Pendray, G.E., eds. (1970) *The Papers of Robert H. Goddard*, New York, p. 23
25 Crouch, T. D. (1999) *Aiming for the Stars: The Dreamers and Doers of the Space Age*, Washington, D.C., p. 366
26 Tsiolkovsky, K. E. (1895) *Grezi o zemle i nebe (Dreams of the Earth and Sky)* Moscow, Isd-vo
27 Benford, G. (1999) Preface to *Last and Last First Men* by Olaf Stapledon, Millennium, p. X
28 Isaacs, L. (1977) *Darwin to Double Helix: the Biological Theme in Science Fiction*, London and Boston, p. 6
29 Henkin, L. J. (1963) *Darwinism in the English Novel 1860–1910*, New York
30 Isaacs, L. (1977) *Darwin to Double Helix: the Biological Theme in Science Fiction*, London and Boston, p. 19
31 Aldiss, B. (1973) *Billion Year Spree*, London, p. 133
32 Stapledon, O. (1930) *Last and First Men*, London (1999) p. 131
33 ibid., p. xiii
34 ibid., pp. xv–xvi
35 ibid., p. xiii
36 ibid., Chapter 16
37 Isaacs, L. (1977) *Darwin to Double Helix: the Biological Theme in Science Fiction*, London and Boston, p. 24
38 Stapledon, O. (1930) *Last and First Men*, London (1999) Chapter 4
39 ibid.
40 Quoted in Isaacs, L. (1977) *Darwin to Double Helix: the Biological Theme in Science Fiction*, London and Boston, p. 47
41 McCarthy, P. A. (1981) Star Maker: Olaf Stapledon's Divine Tragedy, *Science Fiction Studies*, **8**, 25, pp. 266–79
42 Stapledon, O. (1937) *Star Maker*, London (1999) p. 13
43 McCarthy, P. A. (1981) Star Maker: Olaf Stapledon's Divine Tragedy, *Science Fiction Studies*, **8**, 25, pp. 266–79
44 Stapledon, O. (1937) *Star Maker*, London (1999) Chapter 2
45 ibid.

46 Aldiss, B. (1986) *Trillion Year Spree*, Gollancz, p. 197
47 ibid., p. 198
48 Stapledon, O. (1937) *Star Maker*, London (1999) Chapter 1
49 Scholes, R. (1975) *Structural Fabulation: An Essay on Fiction of the Future*, Notre Dame, p. 64
50 Dick, S. J. (1993) Consequences of Success in SETI: Lessons from the History of Science, Progress in the Search for Extraterrestrial Life, 1993 Bioastronomy Symposium, *Conference Proceedings of the Astronomical Society of the Pacific*, **74**, 521–32
51 Shapley, H. (1958) *Of Stars and Men: Human Response to an Expanding Universe*, Boston
52 Struve, O. (1961) *The Universe*, Cambridge
53 Berenzden, R. (1975) in *Copernicus Yesterday and Today*, Vistas in Astronomy, Vol. 17, eds. A. Beer and K. Strand, New York, pp. 65–83
54 Stapledon, O. (1937) *Star Maker*, London (1999) p. 89
55 Haldane, J. B. S. (1934) *Fact and Faith*, London
56 Haldane, J. B. S. (1929) The Origin of Life, *Rationalist Annual*, London
57 Quoted in Dick, S. J. (1996) *Biological Universe*, Cambridge, p. 378
58 Bernal's 1952 lecture to The British Interplanetary Society on November 8 is described in A. E. Slater's *The Evolution of Life in the Universe*, JBIS, May 1953, pp. 114–18.
59 Bernal, J. D. (1951) *The Physical Basis of Life*, London, pp. 14–15
60 Nunan, E. E. and Homer, D. (1981) Science, Science Fiction, and a Radical Science Education, *Science Fiction Studies*, **8**, 25, pp. 311–30

7

Ever since SETI: astrobiology in the space age

Behind every man now alive stand thirty ghosts, for that is the ratio by which the dead outnumber the living. Since the dawn of time, roughly a hundred billion human beings have walked the planet Earth. Now this is an interesting number, for by a curious coincidence there are approximately a hundred billion stars in our local universe, the Milky Way. So for every man who has ever lived, in this Universe there shines a star. But every one of those stars is a sun, often far more brilliant and glorious than the small, nearby star we call the Sun. And many – perhaps most – of those alien suns have planets circling them. So almost certainly there is enough land in the sky to give every member of the human species, back to the first ape-man, his own private, world-sized heaven – or hell. How many of those potential heavens and hells are now inhabited, and by what manner of creatures, we have no way of guessing; the very nearest is a million times farther away than Mars or Venus, those still remote goals of the next generation. But the barriers of distance are crumbling; one day we shall meet our equals, or our masters, among the stars. Men have been slow to face this prospect; some still hope that it may never become a reality. Increasing numbers, however, are asking: 'Why have such meetings not occurred already, since we ourselves are about to venture into space?' Why not, indeed? Here is one possible answer to that very reasonable question. But please remember: this is only a work of fiction. The truth, as always, will be far stranger.

Arthur C. Clarke, *2001: A Space Odyssey*[1]

We don't want to conquer the cosmos, we simply want to extend the boundaries of Earth to the frontiers of the cosmos ... we don't want to enslave other races, we simply want to bequeath our values and take over their heritage in exchange. We are only seeking Man. We don't know what to do with other worlds ... We are searching for an ideal image of our own world: we go in quest of a planet, of a civilization superior to our own but developed on the basis of a prototype of our primeval past.

Stanislaw Lem, *Solaris*[2]

'All right, I'll tell you. But I must warn you that your question, Richard, comes under the heading of xenology. Xenology: an unnatural mix of science fiction and formal logic. It's based on the false premise that human psychology is applicable to extraterrestrial intelligent beings.'
'Why is that false?' Noonan asked.
'Because biologists have already been burned trying to use human psychology on animals. Earth animals, at that.'
'Forgive me, but that's an entirely different matter. We're talking about the psychology of rational beings.'
'Yes. And everything would be fine if we only knew what reason was.'
'Don't we know?' Noonan was surprised.
'Believe it or not, we don't. Usually a trivial definition is used: reason is that part of man's activity that distinguishes him from the animals. You know, an attempt to distinguish the owner from the dog who understands everything but just can't speak. Or how about this hypothetical definition. Reason is a complex type of instinct that has not yet been formed completely. This implies that instinctual behaviour is always purposeful and natural. A million years from now our instinct will have matured and we will stop making the mistakes that are probably integral to reason. And then, if something should change in the universe, we will all become extinct – precisely because we will have forgotten how to make mistakes, that is, to try various approaches not stipulated by an inflexible program of permitted alternatives.'

Arkady and Boris Strugatsky, *Roadside Picnic*[3]

Pioneers of the airwaves

He died without much money to his name, but Nikola Tesla lived a life rich in colour. Widely esteemed as one of the greatest engineers working in America, Tesla was born in 1856, an ethnic Serb in a village that is now part of Croatia. Variously an inventor, engineer, and businessman, Tesla's theories and patents in electromagnetism formed the basis of wireless communication and radio.

Tesla was also at the forefront of alien contact. As early as 1896, he had proposed that radio could be used as a form of contact with extraterrestrial life.[4] And in 1899, while experimenting on atmospheric electricity using a Tesla coil receiver at his Knob Hill laboratory in Colorado, he reported observing repeated (very probably artificial) signals, which Tesla interpreted as being of extraterrestrial origin.

Tesla thought the signals were grouped in ones, twos, threes, and fours, and decided they were coming from Mars, as a Martian interpretation of all things cosmic was very popular at the time. With the benefit of hindsight, lesser scholars have suggested that Tesla detected nothing, and that he was simply misconstruing the new technology.[5] Others proposed that Tesla had been picking up the natural signal of the Jovian plasma torus, a phenomenon caused by the interplay of Jupiter's magnetosphere and the plasma from its planetary moon Io. Nonetheless, other eminent scholars, such as Guglielmo Marconi and Lord Kelvin, also suggested that radio could be used to establish Martian contact in the years following Tesla's advance. Indeed, Marconi even suggested his stations too had picked up possible Martian signals.[6]

The myth of Mars continued. Percival Lowell had published his original Mars book in 1895. His follow-up, *Mars and Its Canals*, came in 1906, though by 1909 better telescopes had nailed the canal theory firmly in its cosmic coffin. Even so, between 21 and 23 August 1924, Mars entered opposition. It was now closer to Earth than any time in a century before or since. In the United States, a National Radio Silence Day was declared. During a thirty-six hour period on the relevant dates, all radios were quiet for five minutes on the hour, every hour. Noted American astronomer David Peck Todd led the programme, and a cryptographer was assigned to decipher any resulting Martian messages.

Then came SETI.

After the war, physicists Phillip Morrison and Giuseppe Cocconi had identified the potential of the microwave region of the electromagnetic spectrum, proposing a set of initial targets for interstellar communication. A year later in 1960, astronomer Frank Drake from Cornell University made the first modern SETI experiment. Christened 'Project Ozma' after the Queen of Oz in L. Frank Baum's fantasy books, Drake employed a radio telescope at Green Bank in West Virginia. With the scope he examined the stars Tau Ceti and Epsilon Eridani close to the 1.42 GHz marker frequency, a region of the radio spectrum known as the 'water hole' due to its propinquity to hydrogen and hydroxy radical spectral lines. Little of interest was found. But it was a start.

The first SETI conference also took place at Green Bank, also in 1961, and funding programmes for an actual scientific search of extraterrestrial intelligence began in earnest. Scientists from the Soviet Union soon took a strong interest, and in the 1960s performed a series of searches in the hope of detecting powerful radio signals from space. Indeed, international cooperation in the field was heightened by its the first definitive book, *Intelligent Life in the Universe* (1966), written by Soviet astronomer Iosif Shklovskii and US astronomer Carl Sagan.

The book was intriguing. Not only did Sagan and Shklovskii provide a scientific introduction to the subject that was soon to be better known as astrobiology, but they also devoted chapters to extraterrestrial contact. Though they stressed their ideas were speculative, Sagan and Shklovskii nevertheless argued that scholars should countenance the possibility that alien contact had occurred during recorded history. Further, they suggested that sub-lightspeed interstellar travel by a sophisticated alien civilisation was a certainty, especially when considered against the developing pace of technological progress in the late 1960s.

Perhaps inspiring the wave of ancient astronaut books that were to prove popular in the 1970s, Sagan and Shklovskii became more daring. They maintained that recurring occurrences of alien visitation to Earth were credible, and that pre-scientific narratives proffer a possible means of describing contact with outsiders. They cited the tale of Oannes, a fishlike being attributed in several rational and independent ancient sources with introducing agriculture, art, and mathematics to the early Sumerians. They urged the global community of scholars to scrutinise ancient documents for further possible instances of paleo-contact.

Solaris

As the scientific world was abuzz with the élan of contact, fiction produced a sceptic to match the hype. As we shall see, Stanislaw Lem's *Solaris* (1961) is ultimately about the problems of contact, and the shortcomings of any potential communication between man and alien. In exploring and probing the ocean planet Solaris from an orbiting research station, the human scientists too are being probed. The planet is sentient, and delves into the minds of the humans who are analysing it. The ocean planet is able to manifest guilty secrets in human form, which each scientist is forced to personally confront.

With a life in writing ahead of him, and his books translated into forty-one languages, selling over twenty-seven million copies, Lem was born in Lvov, Poland, in 1921. His biographical essay, *Chance and Order*, published in *The New Yorker* in 1984, portrays Lem as a child hungry for science, and full of imagination.[7] Motivated by his father's library on anatomy, Lem soon became a child prodigy, creating fantasy worlds, each with its own maps and archives. By his early teens, Lem had performed so well in a school IQ test that he was known as 'the most intelligent child in Poland', although he was not made aware of the accolade until much later.[8]

Described in 1976 by fellow fiction writer Theodore Sturgeon as the most widely read science fiction author on the planet, Lem is perhaps best known for

Solaris, in all its forms: the 1961 novel, along with the two cinematic interpretations, the 1972 adaptation by acclaimed Russian film director Andrei Tarkovsky, and the 2002 remake by US director Steven Soderbergh. And yet, despite its undoubted popularity, there has until 2012 been only one English version of the text, translated in a rather stilted manner, through French. According to Lem, this has led to inaccurate interpretations of Freudian messages in the text: 'One American reviewer made a fatal mistake in that he was unaware of the fact that the idioms of the Polish original are different – hence they do not allow such [Freudian] conclusions'.[9]

Such Freudian nuance notwithstanding, the narrative arc and intent of *Solaris* are apparent enough. The story unfolds onboard an orbiting research station floating above the ocean planet of Solaris. The book's narrator, Dr Kris Kelvin, a psychologist, gets to Solaris after the long journey from Earth. His brief: to gauge the viability of the continued study of the planet, as precious little progress had been reported to Earth. Kelvin finds a cosmic *Mary Celeste*.[10] The station is deserted, save a large but elusive African woman, and a Dr Snow, who behaves peculiarly on meeting, probing Kelvin's identity and even questioning his existence.

As the story unfolds it becomes clear to Kelvin that the crew, him included, are experiencing the materialisation of physical human simulacra, or 'Phi-creatures'. Through these creatures, each scientist is, in turn, confronted with their most painful and repressed thoughts and memories. Witness the 'giant Negress', Dr Gibarian's visitor, who twice appears to Kelvin, and seems to be unaware of the other humans she meets, or simply chooses to ignore them. His own Phi-creature, Rheya, is a simulacrum of his dead wife, whose suicide a decade earlier had left Kelvin wracked with enduring guilt.

Lem uses techniques to help enhance the threatening and alien feel of the Phi-creatures. The scientists onboard the orbiting station probe the makeup of the creatures. Kelvin, along with Drs Sartorius and Snow, attempts an analysis and understanding of *Solaris*, and its associated phenomena. They carry out tests. Kelvin takes and investigates blood samples from the Rheya simulacrum, finding the Phi-creatures to be neutrino-based. The discovery leads eventually to the humans making an anti-neutrino gun, capable of destroying the Phi-creatures, as all other attempts to terminate them had met with failure. Though human in appearance, they seem indestructible. Kelvin had previously tried 'removing' the Rheya simulacrum in one of the escape pods. On her return, and upon realising she is not the real Rheya, the simulacrum unsuccessfully tries suicide by drinking liquid oxygen.

During his stay on the station, and cut off in his cabin with only the Rheya simulacrum for company, Kelvin attempts to understand the ocean planet. He delves into the research station archives, a library holding hundreds of volumes, which represent a century of research into Solaristics. But more is less, for Kelvin. As he ploughs through the archives, the more he realises that Terran science will never understand the alien complexities of the ocean planet and its phenomena. Likewise, the crew continue their attempts at contact with the ocean through X-ray stimulation linked to Kelvin's brain activity. All comes to nought. And Kelvin is left questioning the very idea of contact and the nature of life in the universe, concluding that the ocean planet is a flawed God-like sentience that cannot be understood by human science.

The problems with contact

The moral of Lem's message is clear. The stars are not for us. No matter how much we desire contact and communication with an alien civilisation, it may never be achieved. Terran science may not decipher something so alien from the human, making a symbiotic communication between alien civilisations nigh on impossible. Indeed, Lem also explored the alien contact theme in two other novels, *Eden* (1959), in which there is face-to-face contact with aliens but a barrier to translation, and *His Master's Voice* (1968), where, as with *Solaris*, the barrier of translation is further explored by taking away the possibility of face-to-face contact.

Lem's *Solaris*, then, is a book about the culture of science and its impact on contact. In particular, it focuses on the ideology of scientific culture, which, in the context of Lem's narrative, revolves around the science of Solaristics. The surface of the Solaris ocean is composed of a colloidal substance that can morph into many shapes, some of which ('mimoids') are simulacra of terrestrial objects. But the ocean planet has so long been a subject of study for Terran scientists that Solaristics has evolved into an institution. And so the story centres on the history and controversies of this alien science. The nature of the planet itself, as an example of alien life, and the empirical data gleaned by the research station, also feed into Solaristic studies.

The other obsession in *Solaris* is the question of coding. As we have said, the Phi-creatures (also called Visitors, or 'polytheres') are a kind of neural projection, materialising from the brains of the human scientists onboard the research station. Their origin lies in the most enduring imprints of memory, those that are well rooted, but of which no individual imprint can be singled out. And any attempt to analyse the intent of the occurrences of the Phi-creatures is doomed to failure, blocked by the anthropomorphism of the scientists who 'own' the memories. As Lem would have it, and in Freudian terms, the Solaris ocean has made flesh the forces of the id. And the blocking of the 'owners' represents the hypothesised Freudian mechanisms that function between the id and the ego.

Solaris is a psychic ocean. In the same way that Terran scientists are able to read genetic information, coded into the sequences and structures of the DNA located on chromosomes, the ocean planet is able to understand the physicochemical psychic processes of the brain. In the final analysis, the logic goes, any human explanation of learning and memory ultimately involves some amalgamation of neurons, RNA, and proteins, the chemicals of life through which hereditary information is stored and reused. The Solaris ocean is able to understand and retrieve such stored imprints and makes flesh the psychic tumours of each scientist on the research station.

In this way, Lem uses *Solaris* to focus on two limitations of the culture of science: its anthropocentricism, and its institutionalised nature. These two limitations of science are shown by Lem to produce a disabling mysticism. Kelvin arrives on Solaris with all good intentions, remembering, 'that thrill of wonder which had so often gripped me, and which I had felt as a schoolboy on learning of the existence of Solaris for the first time'.[11] But Lem's narrative traces Kelvin's decline into mysticism. In the end, the despair Kelvin feels is paralleled by the evaporation of his confidence in the explanations of mainstream science. To make matters worse, there is the Rheya simulacrum. The futile relationship Kelvin has with the reincarnated Rheya serves to focus on the subjective side of science. His emotions prevail over his scientific rationality, which accentuates the latter's fragility. Indeed, it is only Rheya's suicide that liberates him once more to focus on the question of Solaris.

Kelvin's is a journey into the dark heart of science. As he delves into the history of Solaristics, he becomes increasingly disillusioned about the ability of science to explain alien phenomena. Lem carefully chooses the character of Kelvin. Through him as narrator, the reader gets a strong sense of the self-defeating journey upon which Kelvin, and through him not just Solaristics but all alien science, is embarked. The field was once vital. But now, all creativity and originality is gone. All that is left is data collation, and the shovelling of new facts into a degenerated and machine-like worldview.

Lem engages the pluralist debate from the perspective of paradigm. At the time of the writing of *Solaris*, American historian and philosopher of science Thomas Kuhn was making his mark with his analysis of the progress of scientific knowledge. In two books, *The Copernican Revolution: Planetary Astronomy in the Development of Western Thought* (1957) and *The Structure of Scientific Revolutions* (1962), Kuhn developed the idea that scientific fields undergo periodic 'paradigm shifts'. When one worldview, or paradigm, gives way to another, all scientific challenges are met from within the boundaries of this new framework. And importantly, the notion of scientific truth, at any time, cannot be established solely by objective criteria, but is defined by a consensus of a scientific community.

In Kuhnian terminology, Solaristics needs a new paradigm. The current contested paradigm of Solaristics flies in the face of the facts implied by the alien phenomena and a suggested alternative paradigm, which is just out of reach. The paradigms are incommensurable; they are competing accounts of reality, which cannot be coherently reconciled. The Terran comprehension of science can never be truly objective. Indeed, it is Gibarian, the most sceptical but optimistic of the scientists, who had come closest to providing this new perspective, in his contact with the ocean.

Geocentrism is a barrier to insight. Though Kelvin understands the limitations of science, it is Snow who has to point out that the geocentrism of science is the root cause. As Snow says, quoted in part at the opening of this chapter:

> We take off into the cosmos, ready for anything: for solitude, for hardship, for exhaustion, death. Modesty forbids us to say so, but there are times when we think pretty well of ourselves. And yet, if we examine it more closely, our enthusiasm turns out to be all a sham. We don't want to conquer the cosmos, we simply want to extend the boundaries of Earth to the frontiers of the cosmos. For us, such and such a planet is as arid as the Sahara, another as frozen as the North Pole, yet another as lush as the Amazon basin. We are humanitarian and chivalrous; we don't want to enslave other races, we simply want to bequeath them our values and take over their heritage in exchange. We think of ourselves as the Knights of the Holy Contact. This is another lie. We are only seeking Man. We have no need of other worlds. A single world, our own, suffices us; but we can't accept it for what it is. We are searching for an ideal image of our own world: we go in quest of a planet, a civilization superior to our own but developed on the basis of a prototype of our primeval past. At the same time, there is something inside us which we don't like to face up to, from which we try to protect ourselves, but which nevertheless remains, since we don't leave Earth in a state of primal innocence. We arrive here as we are in reality, and when the page is turned and that reality is revealed to us – that part of our reality which we would prefer to pass over in silence – then we don't like it anymore.[12]

Terrestrial science is not fit for extraterrestrial exploration. Snow suggests that the inherent limitations of Solaristics, and by extrapolation Lem is referring to all human science, render it unfit for the mission of alien contact and communication. Neither will Snow replace science with the divine. He rejects the proposals of Kelvin to substitute mysticism for methodology in future scientific training courses. Finally, though the other scientists elect to leave the orbiting research station, Kelvin decides to stay. As Kelvin cites his reasons, we catch another glimpse of the contradictions at the heart of the research. Kelvin wishes to remain 'in the faith that the time of cruel miracles was not past',[13] referring to the 'miracles' of contact with the Solaris ocean and the revisit of Rheya, though he fully recognises his desperate and alien situation in this wonderful quote:

> On the surface, I was calm: in secret, without really admitting it, I was waiting for something. Her return? How could I have been waiting for that? We all know that we are material creatures, subject to the laws of physiology and physics, and not even the power of all our feelings combined can defeat those laws. All we can do is detest them. The age-old faith of lovers and poets in the power of love, stronger than death, that finis vitae sed non amoris, is a lie, useless and not even funny. So must one be resigned to being a clock that measures the passage of time, now out of order, now repaired, and whose mechanism generates despair and love as soon as its maker sets it going? Must I go on living here then, among the objects we both had touched, in the air she had breathed? In the name of what? In the hope of her return? I hoped for nothing. And yet I lived in expectation. Since she had gone, that was all that remained. I did not know what achievements, what mockery, even what tortures still awaited me. I knew nothing, and I persisted in the faith that the time of cruel miracles was not past.[14]

Lem is the alien contact sceptic, the lone voice in the wilderness of deep space. *Solaris* is his first true manifesto on the limitations of science. It is a tale in which science is limited not only as a methodology, but also as a faith. And the provenance of the limits of science is rooted in its geocentrism. But *Solaris* is also a despairing novel. It suggests no plausible alternatives. Written at the time of the instauration of SETI, Lem portrays scientists confronted with extraterrestrial phenomena that exceed the limitations of their terrestrial science. Pluralism may have been developed into the theoretical quantitative science of Solaristics, but it cannot account for their actual experience of extraterrestrial contact and communication.

His Master's Voice

Stanislaw Lem was to develop his ideas on extraterrestrial contact in another novel, *His Master's Voice* (1968). The novel takes the form of a memoir, that of a Professor Peter Hogarth, a mathematician who is recruited to work on the His Master's Voice (HMV) Project, part of a government working party of leading scholars set up to decode a 'letter from the stars', an extraterrestrial neutrino beam from somewhere in deep space.

The Project stumbles upon an early finding. In highly concentrated form, coupled with extended exposure, the extraterrestrial beam may act as a fixer to foster the development of life in conditions similar to the early Earth. The Project scholars also find success in using part of the extraterrestrial message to produce a new material. This material is nicknamed 'Frog Eggs' by the biologists, and 'Lord of the Flies' by the physicists. And each team make their samples in separate trials.

The 'Lord of the Flies' material proves to have astounding properties. It can produce energy through cold nuclear reactions. And it can 'transport' this energy instantly to any desired extraterrestrial location – quickly coined the TX effect. Hogarth and his group fear for the consequences: what if this material is to be appropriated by the military to feed their enduring obsession with superweapons? The scholars now continue their work in secret. At last, they discover the material is probably useless as a weapon, as its accuracy diminishes with increasing power. And so they unveil their research conclusions to the Science Council. Quickly, military scientists corroborate their work. And yet, Hogarth and his group are charged with 'antigovernment conspiracy' in initially covering up their findings.

But, as usual, covert military operations had been going on behind closed doors. Once the 'Lord of the Flies' research is made public, it becomes clear that a clandestine military project, dubbed 'His Master's Ghost', had been running in parallel to HMV. Sickened, Hogarth chooses to leave. But he cannot resist the lure of extraterrestrial research. He returns a year later determined to solve the neutrino cypher, but he is fearful of further harmful discoveries.

The closing chapters of Lem's novel showcase a collection of hypotheses on the 'letter from the stars'. Of most interest are those of a cosmogonist, an astrobiologist, and Hogarth himself. From his writings, it is apparent that Hogarth had an epiphany while working on the TX effect. He believes in an extraterrestrial hypothesis, and thinks a 'Sender' lies out in the cosmos, waiting to communicate through the 'letter from the stars'. And if they do not parley with humans, they will merely move on to another, and more worthy, species.

As with *Solaris*, Lem uses *His Master's Voice* to develop his ideas on contact. More specifically, he focuses on the problems associated with communication, and the limits of human intellect. Witness his use of analogy that the HMV Project research was a mere chipping 'a few flecks from the lock that sealed the gate',[15] rather than a deeper insight into the message's meaning. Lem has his protagonist, Hogarth, declare, 'The amount of information that is necessary for even a general grasp of the questions dealt with in the Project exceeds ... the brain capacity of a single individual',[16] and, 'one of the first duties of a scientist is to determine the extent of not the acquired knowledge ... but ... of the ignorance, which is the invisible Atlas beneath that knowledge'.[17] And so Lem's pessimism is clear. His lack of confidence in man's ability to garner understanding through alien contact continues. If some say science has all the solutions, Lem argues that science simply shows how ignorant we remain.

Lem delivers his sceptical verdict on alien–human communication.

At a crucial point in the narrative, Hogarth suggests, 'The message could take the form of numerous things, a letter, a recipe or a manual for building a replica of itself or a particular object'.[18] On the other hand, he posits that an alien tongue might not have the same rules as human languages, the letter being sent, 'without introductions, without a grammar, without a dictionary'.[19] Another constraint might be that of technological prowess, as the 'disparity between levels of civilization'[20] is such that the message recorded by the HMV Project on magnetic tape is merely background noise. The true and latent message lies buried, unrecorded by terrestrial technology.

Cultural difference is another factor. The potential chasm in cultural norms and modes between humans and the 'Senders' could represent an insuperable hurdle to translation, Lem suggests. Hogarth feels it impossible to distil language from cultural context, the enormity of such a task being emphasised by the use of the comparison, 'We are proceeding like a man who looks for a lost thing not everywhere, but only beneath a lighted street lamp, because there it is bright'.[21] And the limits of enquiry itself are once again explored:

> Man's quest for knowledge is an expanding series whose limit is infinity, but philosophy seeks to attain that limit at one blow, by a short circuit providing the certainty of complete and inalterable truth. Science meanwhile advances at its gradual pace, often slowing to a crawl, and for periods it even walks in place, but eventually it reaches the various ultimate trenches dug by philosophical thought, and, quite heedless of the fact that it is not supposed to be able to cross those final barriers to the intellect, goes right on.[22]

The question of random chance is another of Lem's approaches to contact. For one thing, the alien message is uncovered in the first place due to a chance reading of a newspaper found on the seat of a train. For another, Hogarth tells of the pure chance of Earth intercepting the Senders' message, which could simply have 'lay in the path ... between two "conversing" civilizations'.[23] Even blind evolution is viewed as a complexity of culminating chance, a series of contingencies waiting to happen.

One path transpires, a billion others perish. And witness Hogarth's last words in the narrative, as they linger on the inspiration of chance, 'could the whole thing have been only a series of coincidences? Was not the neutrino code itself discovered by accident? And could not the code turn out to have arisen by accident ... and ... by sheer chance produced Lord of the Flies? That is all possible.'[24]

Lem is, of course, also drawing parallels between the HMV Project and the Manhattan Project. Like Manhattan, the atomic bomb research and development programme, carried out by the United States, the United Kingdom, and Canada during World War II, HMV is also a government project, housed on an ex-military desert base, and impacted upon greatly by the military–industrial complex, due to the nature of the research. Lem also penned the novel at the height of the Cold War, and the political context bred conclusions regarding the military repercussions of the research findings proposed in the story. In some ways, Hogarth is similar to Leo Szilárd, the Austro-Hungarian physicist responsible for the creation of the Manhattan Project. As the war unfolded, Szilárd became increasingly troubled that scientists were losing control over their research to the military, and argued many times with the military director of the project. Likewise, Hogarth decides to keep the TX effect under wraps because of his worry that it would be used as a weapon, and refers to the 'global threat' and the 'opposing side',[25] echoing the political climate of the day.

Lem's work was also influenced by his relationship with Soviet politics and culture. Ivan Yefremov, the Soviet paleontologist, social thinker, and science fiction writer, had written a number of key novels in the golden age of Soviet science fiction, widely considered to have been between 1956 and the early 1970s. Yefremov's key novel was the 1957 work *Andromeda: A Space-Age Tale*. Now Soviet writers had often adopted a dialectic double vision. On the surface, they seemed to support official Soviet ideology, optimistic of a communist utopia, set in a vague and distant future. But lurking underneath this veneer, and under the guise of anti-utopias set elsewhere, they deliver a devastating picture of a corrupt, crushing bureaucracy and a police state that exists in the here and now.

In this light it is interesting to consider Lem's echoing of Yefremov's idea that aliens must necessarily have reached a higher level of evolution, that is, the perfect communist state, and to have realised space travel. The assumption follows that this would negate potential threats of conflict between two similarly evolved and sophisticatedly advanced civilisations. Witness also Hogarth's conviction that there was little threat to Earth from the 'Senders' as they must surely possess a high form of intelligence, incapable of conveying any message that could be deemed as harmful. He believed that the production of 'Lord of the Flies' was a 'false reading of the code',[26] fostered by a merely human interpretation of the code, and bearing no relationship to the Senders' true intentions.

And yet, despite Lem's pessimistic outlook on human cognition, commentators have found poise in Lem's work on alien contact, 'Lem balances pessimism with optimism, laughing at the all-too-human idiocy, vanity, and incompetence that have plagued humankind from the beginning'.[27] *His Master's Voice* underlines the fact that Lem believed, 'scientists are, above all, human beings just like the rest of us, with the same kind of passions, open in the same ways to errors or craving for power'.[28]

2001: A Space Odyssey

If there was one writer who was catalytic in developing the alien for the mass market, it was Arthur C. Clarke. As we have seen from a long and illustrious history of imaginative and speculative fiction, from Lucian to Lem, the artistic sweep of the alien motif in fiction was impressive enough. But as far as the propagation of the idea of life in the universe was concerned, it was the sway of cinema that beamed the broadcast even further. Cosmic fiction and its exploration of humanity could be examined without the alien. But the inclusion of the alien was pluralism's defining moment in the public imagination. In addition, science fiction would have stayed marginalised were it not for the opening up of the genre to film and television.

Arthur C. Clarke and Stanley Kubrick's *2001: A Space Odyssey* (1968) was delivered during the infamous extraterrestrial hypothesis, which peaked between 1966 and 1969.[29] The hypothesis held that UFOs were close encounters with visiting aliens, a hypothesis vastly influenced by the mythic fiction of Wells, Stapledon, and Clarke himself.

Celebrated for the maturity of its portrayal of mysterious, existential, and elusive aliens, *2001* raised science fiction cinema to a new level. Eminent US film critic Roger Ebert, when asked which films would remain familiar to audiences 200 years from now, selected *2001*. Another critic claimed the picture was an 'epochal achievement of cinema' and 'a technical masterpiece'.[30] The film, not the book, made Clarke the most popular science fiction writer in the world. Kubrick's masterpiece, which made dramatic and sophisticated use of the pluralist premise, quickly became a classic discussed by many, if not understood by all.

Kubrick's bible was *Intelligent Life in the Universe*, which as we have mentioned was written in 1966 by Carl Sagan and Russian astrophysicist Iosif Shklovskii.

It was the first seminal scientific text on the question of extraterrestrial intelligence. Kubrick had originally filmed interviews with twenty-one leading scientists about their stance on pluralism to act as a prologue to the film's main narrative. Interviewees included physicists Frank Drake and Freeman Dyson, anthropologist Margaret Mead, roboticist Marvin Minsky, and Alexander Oparin, the great Soviet authority on the origin of life, sometimes described as the 'Darwin of the twentieth century'. Kubrick's aim was to lend astrobiology that special dignity it has only acquired since. Though the interviews were cut from the final version of the film, a book of the transcripts was published in 2005.

The Ultimate Trip

2001 is an epic journey, the 'Ultimate Trip', as it was billed in those new-age counter-culture days. Darwin and Wallace's work had inspired German philosopher Friedrich Nietzsche to write *Also Sprach Zarathustra*. The book identifies three stages in the evolution of man: ape, modern man, and, ultimately, superman. As Nietzsche put it, 'What is the ape to man? A laughingstock, or painful embarrassment. And man shall be to the superman: a laughingstock or a painful embarrassment.'[31] Modern man is merely a link between ape and superman. For the superman to evolve, man's will, 'a will to procreate, or a drive to an end, to something higher and farther',[32] must power the change.

In the same way, Kubrick's motion picture traces man's journey through three stages. As the movie's subtitle suggests, the narrative is a spatial odyssey from the subhuman ape to the post-human starchild. The unfolding four-million-year filmic story embraces each theme of science fiction: space (contact through alien cultural artifacts), time (evolutionary fable), machine (the man-machine encounter with HAL, computer turned murderer), and monster (human metamorphosis).[33] The opening 'Dawn of Man' scene of *2001* sees the Sun rise above the primeval plains of Earth, to the rising soundtrack of Richard Strauss' Nietzsche-inspired tone poem, *Also Sprach Zarathustra*. A small band of man-apes are on the long, pathetic road to racial extinction.

The journey begins with one of the hominids proudly hurling an animal bone into the air. In an astounding cinematic ellipsis the bone instantly morphs into an orbiting satellite, and three million years of hominid evolution is written off in one frame of film. The agency that drives the guided evolution of these early hominids is an alien artifact in the shape of a black monolith. Like Wells' Martians, the monolith embodies the void.[34] Primal bone technology marks the birth of the modern era. Man and machine, from the very outset, are inseparable.

The mysterious presence of the monolith transforms the hominid horizon. The journey to superman begins.

The ingenious use of film technology is one of the key factors that makes *2001* such a tour de force. Winner of an Oscar for special effects, the movie seemed to offer a more 'realistic' picture of space travel than the endeavours of Armstrong and Aldrin only a year later. Sections of the picture were used in training NASA astronauts. Indeed, Arthur C. Clarke later suggested that of all the responses to the film, the one he valued most highly was that of the Soviet cosmonaut Alexei Leonov, 'Now I feel I've been into space *twice*!'.[35]

With the brooding presence of the alien throughout, humanity's growing maturity through an early space age now unfolds in a three-way narrative of machine, human, and posthuman. A space voyage leads to further way stations of the monolith. Another alien artifact in the shape of a black obelisk is uncovered at a pioneering Moon base. A mission is sent to Jupiter, to where the mysterious lunar artifact seems to be sending its alien signal. The banality and vacuity of the human crewmembers is sharply and ironically contrasted by the robust intelligence of the ship's onboard computer, HAL 9000. The film was one of the very first to carry 'product placements' for companies such as IBM, Pan Am, and AT&T. Indeed, space travel is replete with corporate logos and trademarks, showing a world

> absolutely managed – the force controlling it discreetly advertised by the US flag with which the scientist [Doctor Floyd] often shares the frame throughout his 'excellent speech' ... and also by the corporate logos – Hilton, Howard Johnson, Bell – that appear throughout the space station. In 1968, the prospect of such total management seemed sinister – a patent circumvention of democracy.[36]

The irony of the picture's portrayal of an insipid future dominated by corporations and technology was lost on some. British scientist and software pioneer Stephen Wolfram said the film's futuristic technology impressed him greatly as a boy. And Microsoft mogul Bill Gates has suggested that *2001* inspired his vision of the potential of computers (though whether Gates was also inspired by the portrayal of sinister corporate domination is mere speculation).[37] Nevertheless, such corporate control is symptomatic of the spiritual crisis of the early space age portrayed in the picture.

The potent evolutionary force transmitted by the alien obelisks is overdue. The space age was originally inspired out of the apes by the alien intelligence. At last, the odyssey of self-discovery culminates under the watchful presence of the alien monoliths when modern man, in the form of the individual astronaut David Bowman, comes to an end. With the movie screen replete with the massive presence of planet Earth, the foetus of the superhuman starchild floats into view. Like the forms of man invoked by Stapledon, the starchild moves through space-time without artifice, the image suggesting a new power. Man has transcended all earthly limitations.

A 'scientific definition of God'

American auteur Stanley Kubrick claimed the picture provided a 'scientific definition of God'.[38] There was little drama in the evolution proposed by Wallace and Darwin, just the slow, solid state of inexorable change. So Kubrick and Clarke conjured up a fictional form of Stephen J. Gould's 'punctuated equilibrium'. The movie augments the usual driving force of evolution, long periods of steady change, with the episodic guiding hand of superior alien beings. It is a story of the effective creation and resurrection of humanity.

Clarke had toyed with such ideas before. His 1953 novel *Childhood's End* was an archetype of the way in which alien fiction developed the idea of the post-human. And Clarke had already written a number of short stories on the alien motif. The influential *The City and the Stars* portrays humanity confronted with extraterrestrial cultures and intelligences: 'He could understand but not match, and here and there he encountered minds which would soon have passed altogether beyond his comprehension'.[39]

In *Childhood's End* Clarke had developed the myth of contact through alien invasion. The story's 'Overlords' are benevolently responsible for guiding humanity to an even greater intelligence, the 'Overmind'. Clarke uses the alien context to highlight humanity's immaturity in an ancient and cultured universe. The Overlords exact an end to poverty, ignorance, war, and government. But there is a price to pay. It is a preparation for the final destiny of humanity; and Earth children are to be sacrificed, and united within the collective of the Overmind.

Clarke viewed the question of pluralism in a very positive light: 'The idea that we are the only intelligent creatures in a cosmos of a hundred billion galaxies is so preposterous that there are very few astronomers today who would take it seriously. It is safest to assume, therefore, that They are out there and to consider the manner in which this fact may impinge upon human society.'[40] *Childhood's End* was written midst growing claims for inexhaustible exoplanetary systems, though it was not until 1995 that empirical evidence for such exoplanets was finally discovered.

Much of Clarke's fiction reflects his scientific belief in extraterrestrial life, and eventual contact. Amusingly, in the preface of a 1990 reprint and partial rewrite of *Childhood's End*, Clarke attempts to unravel pseudoscience from the extraterrestrial message underlining the original narrative: 'I would be greatly distressed if this book contributed still further to the seduction of the gullible, now cynically exploited by all the media. Bookstores, news-stands and airwaves are all polluted with mind-rotting bilge about UFOs, psychic powers, astrology, pyramid energies.'[41]

But Clarke was adamant about the continued relevance of pluralist fiction, such as *Childhood's End* and *2001*: 'I have little doubt that the universe is teeming with life. SETI is now a fully accepted department of astronomy. The fact that it is still a science without a subject should be neither surprising nor disappointing. It is only within half a human lifetime that we have possessed the technology to listen to the stars.'[42]

As the most recent case of Clarke will testify, the influence of fiction on pluralism has been huge. Physical scientists have historically held a deterministic view of the possibility of extraterrestrial life. And as the Clarke quotes above show, this determinism is based mostly on the physical forces in the universe. The idea that the sheer number of stars and orbiting planets is somehow statistically sufficient to suggest that other Earths lie waiting in the vastness of deep space. Fiction, for many centuries, followed suit. Since Copernicus came before Darwin, and physics before biology, fictional accounts of alien life have usually been positioned firmly in the pro-SETI, pro-life camp of the pluralist debate. By the twentieth century, an entire generation of future SETI-hunters was cast under the same spell.

But as the century progressed, the story changed. Two of the main pioneers of the evolutionary synthesis, that fusion of evolutionary biology with genetics, were Theo Dobzhansky and Ernst Mayr. These scholars emphasised that whilst physics and fiction still think along deterministic lines, evolutionists are impressed by the incredible improbability of intelligent life ever to have evolved, even on Earth. Or, to put it in the powerful words of American anthropologist Loren Eisley:

> So deep is the conviction that there must be life out there beyond the dark, one thinks that if they are more advanced than ourselves they may come across space at any moment, perhaps in our generation. Later, contemplating the infinity of time, one wonders if perchance their messages came long ago, hurtling into the swamp muck of the steaming coal forests, the bright projectile clambered over by hissing reptiles, and the delicate instruments running mindlessly down with no report . . . in the nature of life and in the principles of evolution we have had our answer. Of men elsewhere, and beyond, there will be none forever.[43]

We may, after all, be alone in the universe. But, such has been the power of science fiction that its exploration of evolution and the future of man led directly to a huge investment in the serious search for ETI.

Rendezvous with Rama

By now, fiction had been imagining the alien for many centuries. And as, in all that time, science still had little to say about the actual details of extraterrestrial life, some had turned to more elusive explorations of alien contact. In the wake of Kubrick and Clarke's *2001*, two fascinating narratives on the question of alien life were written in the early 1970s. One was *Roadside Picnic*, written in 1972 by Soviet-Russian science fiction writers Arkady and Boris Strugatsky. The other, which we shall look at first, is *Rendezvous with Rama*, a novel penned by Clarke himself in 1973.

In the wake of *2001*, Arthur C. Clarke had signed a three-book deal, something of a record for a science fiction author at the time. The first of these novels was *Rendezvous with Rama*, which won Clarke all the major genre awards, spawning an entire series and gifting him, along with the *2001* sequels, the backbone of a career that was to last the rest of his life.

Clarke and Kubrick's elusive alien returned to the Solar System.

In Clarke's novel, Rama is the terrestrial name for what turns out to be a rogue spaceship, but which was first mistaken for an asteroid, and so named for the king considered to be the seventh avatar of Vishnu, a Hindu god. The unidentified flying object is initially christened asteroid 31/439, and detected in the year 2130. Spotted in the vicinity of Jupiter, the object's calculated speed is over one hundred thousand kilometres an hour.

But the object's trajectory tells another tale. This is no long-period orbit, originating from the outer Solar System and fetching the Sun as a compass. This object stems from interstellar space. Not only that but astronomers are fascinated to find that the object has a rapid rotational period of four minutes. A space probe is sent to investigate. Launched from Phobos, the Martian moon, the probe makes an earth-shattering discovery. A rapid fly-by reveals a perfectly cylindrical spaceship, measuring thirty-four miles by twelve, and made from some mysterious, flawless material.

It is man's first encounter with an alien spaceship.

A manned mission is sent to scrutinise the vessel further, inside as well as out. A survey ship, *Endeavour*, is closest at hand to catch Rama during its brief spell in the Solar System. And so, one month after it first came to Earth's attention, the mysterious Rama is unveiled. With painstaking care and a wonderful sense of suspense, Clarke shows better than ever in *Rama* his celebrated skill at conveying the awe of scientific discovery. With the giant alien ship already within the orbit of Venus, the twenty-plus crew unlock the long cylindrical microcosm of alien culture.

Clarke's poise is perfect. Throughout the novel, the alien nature and intent of the spaceship, and its owners, remains elusive. The terrestrial explorers uncover an ecosystem. Just as the Sumerian cylinder seals from Earth's own ancient history tell a picture story on a geometrical surface, so the architectural evidence created within Rama's built environment speaks of an extraterrestrial culture, alien to man. The astronauts find several key features. They discover what appear to be cities, blocked building-like structures, delineated by 'streets', and with shallow grooves, which may act as vehicular trolley tracks. The Cylindrical Sea is discovered, a banded ocean that spans across the inner surface of Rama, cutting off the human explorers from the far end of the cylindrical world.

At the far southern end of Rama are seven gargantuan cones, thought to be the propulsion system with which Rama travels through interstellar space. At first it seems the cones are beyond exploration, due to the Cylindrical Sea and the small matter of a five-hundred-metre cliff on the opposite shore. But one of the crewmembers braves the journey on a skybike, finding that the massive metal cones generate magnetic fields of flux, and lightning bolts take out his bike.

On regaining consciousness from a bike crash created by the bolts, our cone explorer encounters the first signs of sentient life. A crab-like creature is dicing up his bike, as if in agreement with some cosmic recycling scheme. But it is unclear whether the creature is some kind of service drone, or a biological alien. As the creature dumps the diced remains of the skybike into the sea, our cone explorer approaches the creature with care. He is totally ignored. A rescue party sails the Cylindrical Sea in a makeshift craft to pick him up, as Clarke treats us to a description of the tidal sea, made turbulent by a change of trajectory by Rama itself.

When the crew return to base, they find peculiar creatures examining their base camp. And when one of the creatures accidentally plunges off a Raman walkway to its destruction, the crew biologist is able to carry out a post-mortem. The creatures are 'biots', neither biological life-form, nor robot, but both, a carefully engineered hybrid. The crew believe the biots to be service drones of the Raman starship and watch them with increasing fascination. As Rama draws close to perihelion, the biots act more strangely still. Like the lemmings of legend, they leap into the Cylindrical Sea.

Soon the crew must leave. One last idea emerges. Some crewmembers resolve to explore a city (nicknamed 'London'), chosen for its proximity to the escape stairway at the northern end of the cylindrical world. They cut into one of the city buildings using laser technology, and stare open-mouthed at what they uncover. The building is a hologram store, a housing replete with holographic images that are thought to have some important function for the mysterious Ramans. They catch a tantalising glimpse of a Raman uniform moments before they are forced to leave this world behind.

Rama leaves the human world in its wake. From the relative safety of their survey ship *Endeavour*, the crew look on as Rama slingshots its way out of the Solar System in the vague direction of the depths of the Large Magellanic Cloud, the small satellite galaxy of our own Milky Way. Somehow channeling our Sun's gravitational field with its enigmatic 'space drive', Rama accelerates and is seen no more.

Clarke's *Rendezvous with Rama* is a study in science exposition. There are no aliens, and the characterisation is so slim even the humans are barely recognisable. And yet Clarke's account is remarkable and often magical. The story itself is rather simple. A few humans board a ship, explore its contents, but are forced to leave, just before the ship disappears forever, caring little for man and his curiosity. But the reading experience is very different. Clarke's creative and imaginative style engages and captivates the reader with this colossal artifact of alien culture. The fine points of future technology are laid out before us. And Clarke is able to deliver the entire concept as one package, a gift to be unwrapped, layer after layer, wonder after wonder. This technological future maybe our future too, these alien artifacts may one day be within the scope of man.

The rendering of alien culture in Rama is 'as magical as it is mysterious'.[44] Writing in *The New York Times*, critic John Leonard suggested that, though Clarke was cold and indifferent to the 'niceties of characterization',[45] the novel was outstanding for its conveying 'that chilling touch of the alien, the not-quite-knowable, that distinguishes sci-fi at its most technically imaginative'.[46] Fellow British science fiction writer Brian Aldiss has this to say about Clarke and Rama in his history of science fiction, *Trillion Year Spree*:

> Towards the end of 'A Meeting with Medusa' our attention is brought to bear on a sign which reads ASTONISH ME. This seems to be Clarke's credo. At his finest he lives up to it, and has done so for over forty years. His imagination works best when dealing with the abstractly scientific or philosophical. Like certain painters he feels uncomfortable drawing the human form. His brand of mystical realism works best when the conceptual heart of his fiction is of the scale of a Medusa or a Rama.[47]

The Brothers Strugatsky

Stanislaw Lem was keenly aware of the continuing question of alien portrayal when, in 1983, he wrote an essay about the Strugatskys' *Roadside Picnic*:

> The strategy theologians apply to their principal subject is not properly available to the writer of science fiction. The mystery of the

> Alien, unlike that of God, cannot be preserved by resorting to dogmatically imposed contradictions without betraying the true nature of science fiction. Yet presenting the Alien has its problems. H. G. Wells' approach in making his Martians physically hideous left them mentally and socially unreconstructed; their motives for invading Earth remain recognizable caricatures of human thinking and hence compromise their Otherness. However, the legion of imitators who have debased the example of *The War of the Worlds* in trying to outdo it in the realm of monstrosity have deposed of that problem by neglecting to furnish their Cosmic Invaders with any motive whatever only to supply themselves with another, by substituting a malign, inverted fairy-tale universe for the real world that science fiction should model itself after. The best way out of such difficulties lies with the method the Strugatskys adopt in *Roadside Picnic*: of not-depicting the Alien.[48]

Arkady and Boris Strugatsky are the most prominent Soviet science fiction writers whose work is well known in the West. Their early work, beginning in the late 1950s, was influenced by Ivan Yefremov. But the book that brought them fame was *Roadside Picnic*, translated into English in 1977, and filmed as *Stalker* in 1979, by Andrei Tarkovsky, the Russian film director who also created the acclaimed cinematic dramatisation of Lem's *Solaris* in 1972.

Roadside Picnic is an alien visitation tale with a difference. The story is set in a post-visitation world, planet Earth, where there are now six mysterious Zones, regions of our globe that have been touched in some way by an alien visitation, some ten years past. The Visitors were never seen. But people local to the Zones reported loud explosions that blinded some and caused others to catch a kind of plague. Though the visit is thought to have been brief, about twelve to twenty-four hours, the half-dozen Zones are full of mysterious phenomena, and where strange events continue to occur.

Deep in the Zones, the laws of physics break down. Some Zone districts earn ominous names from studying scientists. The 'First Blind Quarter', 'Plague Quarter', and 'Second Blind Quarter' – all based on effects the Visitation has had on the local population. The six Zones become contaminated with fatal phenomena, littered with mysterious objects, or artifacts, whose various properties and original intent is so incomprehensible and advanced it might as well be supernatural. The Visitation Zones vary in size, some located in populated towns, most perhaps the size of a few square miles, beset with abandoned buildings, railways, and cars, some decaying slowly, while others seem brand new. The Zones are deadly to all forms of life. They harbour spacetime anomalies, and random spots capable of killing by fire, lightning, gravity, or in other bizarre and alien ways.

The location of the six Zones is not random. This discovery, made by Nobel Prize winning physicist, Dr Valentine Pilman, is explained in a radio interview at the beginning of the book, 'Imagine that you spin a huge globe and you start firing bullets into it. The bullet holes would lie on the surface in a smooth curve. The whole point (is that) all six Visitation Zones are situated on the surface of our planet as though someone had taken six shots at Earth from a pistol located somewhere along the Earth–Deneb line. Deneb is the alpha star in Cygnus.'[49]

The governments of the world, coordinated by the United Nations, try to keep a lid on the Zones. They insist on tight control, trying to stem the flow of alien artifacts out of the Zones, in fear of the unpredictable consequences of such poorly understood alien technology. It is quite possible that a single artifact harbours sufficient power to cause a plague, permanently damage, or even destroy the planet. Pitted against this authoritarian control is a frontier culture of stalkers who inhabit the Zone perimeters. Communities of such stalkers grow around the Zones, a kinship of thieves whose aim is to steal into the forbidden regions and find any potentially lucrative alien artifacts. Stalkers work at night, as by day soldiers and scientists constantly observe the Zones. The story is set in the town of a fictitious Commonwealth country, and tells the tale of one such stalker, over a period of eight years.

At the book's opening, Dr Valentine Pilman makes clear the book's title in his description of an alien Visitation as

> A picnic. Picture a forest, a country road, a meadow. Cars drive off the country road into the meadow, a group of young people get out carrying bottles, baskets of food, transistor radios, and cameras. They light fires, pitch tents, turn on the music. In the morning they leave. The animals, birds, and insects that watched in horror through the long night creep out from their hiding places. And what do they see? Old spark plugs and old filters strewn around ... Rags, burnt-out bulbs, and a monkey wrench left behind ... And of course, the usual mess – apple cores, candy wrappers, charred remains of the campfire, cans, bottles, somebody's handkerchief, somebody's penknife, torn newspapers, coins, faded flowers picked in another meadow.[50]

The tension between the clampdown and the stalkers is told through Redrick Schuhart, an experienced stalker who makes regular and illegal night visits into the Zones. By day, Redrick is employed as a lab assistant at the International Institute, the local agency responsible for studying the Zone. Redrick pays a high personal cost for his excursions into the alien Zones. After an official expedition into the Zone to recover a unique artifact (a 'full' empty[51]), Redrick's friend dies. And when Redrick decides to marry his pregnant girlfriend, he is aware that stalkers carry a high risk of mutations in their children.

On birth, Redrick's daughter appears to be a beautiful, happy, and intelligent normal child, save for the presence of profuse short and light body hair. She is christened, lovingly, Monkey. But soon Redrick's dead father returns home from the cemetery, now sited inside the Zone. Simulacra of other deceased are also slowly returning to their homes. As she grows, Monkey becomes increasingly ape-like, her character more reclusive and less talkative, but strangely screaming at night, together with Redrick's father. Later in the book, medical examinations show that Redrick's daughter is no longer human.

The authorities fight a losing battle with human curiosity and endeavour. An old friend of Redrick's, another Institute worker by the name of Richard Noonan, turns out to be a leading light in a covert governmental taskforce, charged with the seemingly impossible mission of stopping the illicit flow of artifacts from the Zone. Noonan is delusional. Convinced he has achieved his multi-year mission, he is soon confronted by a superior in a clandestine meeting, who reveals that the flow of artifacts is stronger than ever. The clampdown begins anew. Noonan is tasked with finding the culprits, and their covert operation.

The lure of the alien artifacts is hard to resist. Though many are lethal, some have beneficial powers, such as the 'so-so', a round black stick that produces endless energy and is used to power vehicles. Others, such as the 'Death Lamp', emit deadly rays that destroy all life in the vicinity. The holy grail of the alien artifacts is the legendary 'Golden Sphere'. Rumoured to have enough latent power to make any wish come true, the Sphere is buried so deep in the Zone that only one stalker knows the true path to its location.

Another stalker key to the narrative is Burbridge the Buzzard. Redrick helps Burbridge out of the Zone after Burbridge loses his legs to a substance known as 'witches' jelly'. Referred to by Institute scholars as a colloidal gas, the jelly penetrates most known materials. All it touches merely transforms into more witches' jelly. Only certain ceramic vessels are able to contain it. The jelly seems to collect in low-lying areas, such as basements, but is highly volatile. Burbridge sets up 'Sunday School', a weekend 'picnics-for-tourists' business, which is merely a cover for a new generation of stalkers to learn the key skills and strategies of stalking.

Redrick goes into the Zone one last time. His mission is to find the alien wish-granting 'Golden Sphere'. Redrick has a map, gifted to him by Burbridge, whose son joins Redrick on the expedition. But Redrick suspects one of them will have to die in order to reach the sphere, so that a phenomenon known as 'meat-grinder' is disengaged. Redrick keeps this a secret from his young escort. At first, it seems as though their mission is complete. The lure of success brings them to the secret location of the Sphere. After surviving so many obstacles on their difficult journey, Redrick's escort is so inspired that he rushes toward the Sphere, only to be brutally assassinated by the meat-grinder. Exhausted and devastated, Redrick contemplates his broken life. As he strains to find meaning in his past, he is nonetheless hopeful that the Sphere will seek and find something good in his heart, as the legend has it that the Sphere grants hidden wishes. In the end, however, Redrick can think of nothing other than parroting the dead escort's words, 'Happiness For Everybody, Free, And No One Will Go Away Unsatisfied!'.

Science fiction and the elusive alien

Whither the alien?

As we have seen, sampled throughout this book, imagined forms of alien life have been an inexhaustible topic of fantastic fiction for two millennia. At the heart of this literature has been the desire to portray the true alien. We have seen the ways in which writers have tried to depict beings that are blessed with reason, but are not human. The strategy that theologians have adopted, in their portrayal of the equally elusive being of God, has been to preserve a mystery before which reason may be forced into temporary silence. And yet the same approach is limited in science fiction as the genre is expected to work with the facts. After all, aliens are, like us, material lifeforms.

And so a prosaic line of attack has been engaged. Extraterrestrials differ from us physiologically, and from this aspect alone have their characteristics been drawn. Psychologically, it seems, they are little different from humans, as there can only be one form of reason. H.G. Wells first spawned this notion over a century ago in *The War of The Worlds*. Wells' Martians are a horror to behold. And yet, man may one day share their physical form, a future of atrophied body and cranial expansion. Wells seemed to have little to say about the alien nature of Martian culture. Like their atrophied forms, perhaps their society too had merely withered away, a fused future in which both physiology and culture are reduced to simple forms.

Wells can easily be forgiven. He was writing at the close of the century before last. And Wells' narrative is warranted: a parched Martian civilisation, perched on a dying planet that is slowly turning into a desert, makes war for water. A bountiful Earth looms before them, temptation made flesh. Given this context, the behaviour of the Martians is a cold and unsympathetic calculation, rational *in extremis*.

But this rather specific scenario was plagiarised *ad nauseam*, as science fiction stories trotted out mechanical simulacra of the Wellsian prototype, a farcical parade of alien monstrosities, stretching credulity to breaking point. Indeed, the cumulative impact of this genre stereotype was to give the impression of a cosmos replete with regimes whose sole desire to expand seemed wholly irrational. The greater the attributed power of the aliens, the more irrational seemed their invasion of a relatively lowly Earth. Each narrative disguised a kind of paranoid geocentrism, and perhaps in many cases an echoed xenophobic fear of Soviet expansionism.

Geocentrism became rare in fact, but familiar in fiction. Aliens were portrayed with the shallowest of intellects. Whereas Wells' Martians, inhabitants of a small desert world, may well have considered a moist Earth a fitting treasure, these later aliens invade Earth for fun. And so Wells' interplanetary Darwinism was exchanged for cosmic sadism. Any civilisation developed beyond the wheel stage seemed intent on invading Earth, an insignificant planet in the unfashionable arm of a little known spiral galaxy. Some science fiction authors, it seems, merely projected their national fears onto deep sky.

So, as SETI began to blossom, the image of the alien had begun to degrade. At a time when scientists started to seriously discuss the challenge of interstellar communication with alien species, much of fiction could think no further than the latest bug-eyed monsters travelling to the ends of the Galaxy, seeking Earthlings to swat. Xeroxed copies of prototyped alien civilisations were projected into the future from the old days of imperialism. The aliens were colonialist adventurers, plundering privateers on the make, or plotting conquistadors out for imperial invasion. These fairy-tales seemed totally uninformed by the latest findings of science, and the imagined aliens occupied a counterfeit cosmos, one bereft of difference in spacetime and evolution, where all astronomy, physics, and even sociology and psychology were reduced down to geocentric and colonialist copy.

Enter the elusive alien.

The intelligent writer was one who recognised the scientific difference between alien and human, unlike in psychology as well as physiology and form. Witness Clarke and Kubrick's aliens in *2001*. In his book *Cosmic Connection: An Extraterrestrial Perspective*, American astronomer and SETI pioneer Carl Sagan confessed that Kubrick and Clarke approached him on the best way to depict extraterrestrial intelligence in the movie. (You will recall that Sagan's *Intelligent Life in the Universe* was something of a bible for Kubrick, as the film was unfolding from concept into reality.)

Sagan's response is telling. He understood Kubrick's desire to portray aliens as humanoid. It was convenient, after all, and hardly uncommon. But Sagan pointed out that, since alien life-forms were unlikely to bear any resemblance to earthly life, to portray them as such would introduce 'at least an element of falseness',[52] into the film. Rather, Sagan suggested, the film should depict extraterrestrial super-intelligence, fitting for the Nietzschean theme of man's evolving into post-human superman. On attending the film's premiere, Sagan was 'pleased to see that I had been of some help'.[53] And when pressed in an interview with *Playboy* in 1968, Kubrick hinted at the nature of the elusive aliens in *2001* by suggesting, given their long maturation, they had evolved from biological beings into 'immortal machine entities', and then into 'beings of pure energy and spirit', beings with 'limitless capabilities and ungraspable intelligence'.[54] Stapledon's influence on Clarke and Kubrick seems evident here.

Indeed, Sagan also chose an elusive rendering of extraterrestrial super-intelligence in his 1985 novel *Contact*. The book, which originated as a screenplay in 1979 and eventually made it into a successful movie in 1997, is based on the story of a SETI scientist, part of a project team that unveils an alien message, coming from a star system around Vega. It is the first confirmed communication from extraterrestrial beings. At length, once the message is deciphered, the project team unearths plans to build a highly advanced vehicle, capable of transporting passengers through wormholes in spacetime.

But the aliens remain hidden. Perhaps in a knowing nod toward *Solaris*, when five terrestrials travel through spacetime to meet with the message makers they do not encounter their true alien form. Rather, the message senders appear in the guise of persons significant in the lives of the travellers, whether living or dead. In the case of the lead female protagonist, Ellie Arroway, she finds herself in a surreal beachscape, similar to a picture she drew as a child. A fuzzy figure approaches, and materialises into her late father as it comes closer. But she soon understands the father figure is an alien taking a familiar form. The alien deflects her many questions, and simply explains this journey as merely man's first step to joining other spacefaring species: 'You're an interesting species, an interesting mix. You're capable of such beautiful dreams and such horrible nightmares. You feel so lost, so cut off, so alone, only you're not. See, in all our searching, the only thing we've found that makes the emptiness bearable is each other.'[55]

Witness also Lem's *Solaris*. Lem rejected the all-too-easy temptation of an Earth invasion. Instead he chose to limit himself to portraying phenomena that differed markedly from what readers of alien fiction were familiar with. An interventionist alien race, bent on invasion, suggests a cosmic undertaking. But what is their aim? Do they board fleet and nimble silver starships just to scrap, or thieve, on planet Earth? Or do they come just to observe, or for some other puerile reason? The very scenario beggars belief and exposes the significant shortcomings of such an alien portrayal. And so many texts of the twentieth century were exemplars of how *not* to depict an imagined alien civilisation.

Roadside Picnic and the elusive alien

In contrast we have *Roadside Picnic*. The Strugatsky brothers have used the elusive alien to such excellent effect that the text warrants closer examination. The book is founded on two ideas. The first is the strategy of concealing the identity of the alien visitors. In *Roadside Picnic* their mystery is apparent. Not only do we not know what they look like, what they want, or why they came to this humble world, we are not even certain they have been at all. Assuming they have visited, we are not sure why their visit was so brief, or whether indeed they have actually left. They may still be among us. The second idea is the human reaction to the Visitation. There is little doubt that some form of alien culture has visited the Earth, or at least fallen from the sky. City people go blind, become ill with a 'plague'. Deserted areas of Harmont city become one of the Zones, in which the properties of physics seem to break down, distancing it from the human world.

The world keeps on turning. And yet in the Zones there is breakdown. Even when those who populate the Zones escape, the question of causality is warped: ninety per cent of the clients of just one businessman die in the following year of 'natural' causes. We do not learn much, if anything, about the Visitation itself. We find it hard to fit the physics into any of our taxonomy schemes, and this is quite deliberate. Even if human scientists came close to the mechanics of the anomalous events in the Zone, they would be no closer to the heart of the matter, namely the nature of the alien visitors. The technology of the visitors sheds little light on their extraterrestrial culture or makes them appear any less alien. A further comparison with Wells' *War of the Worlds* highlights the difference of *Roadside Picnic* still further.

In Wells' book there is little doubt as to the identity and intention of the alien visitors. As Wells intended, this narrative is in part a broadside against the atrocity of imperial invasion. At the time of publication, aggressive British imperialism had colonised huge swathes of the globe in Africa and Australia, South America and Asia. The readers were presented with Wells' vision of an adversary so totally superior to this British Empire that there is a nightmarish, fundamental breakdown of their contemporary world. There is a collapse of so-called civilised order under the huge blows from Martian technology, and yet Wells asks, 'before we judge them [the Martians] too harshly, we must remember what ruthless and utter destruction our own species has wrought ... The Tasmanians, in spite of their human likeness, were entirely swept out of existence in a war of extermination waged by European immigrants ... Are we such apostles of mercy as to complain if the Martians warred in the same spirit?'[56]

This Wellsian alien invasion has little in common with *Roadside Picnic*. If pressed, we may admit that some kind of invasion has occurred. The fallout Zones are unmistakable and yet the human world beyond those districts goes on as before. Within the Zones, miraculous and alien phenomena occur. But the Strugatskys preserve the identity of their alien culture at all costs. Mere rumour alerts us to the fact that scientific experiments are being carried out on artifacts found in the Zones. Shady, Kafkaesque institutes for the study of extraterrestrial culture are secretly trying to comprehend the nature of the Visitations, but we learn nothing of the impact of this alien culture on world affairs. And this too is a striking difference.

Roadside Picnic is an alien invasion seen from below. The conventional view is the one from above. In tales featuring contact or a visitation from an alien race we are usually met with the scientific and extra-scientific controversy that the contact triggered. Here we have nothing. No news either of novel modes of thought, no transformations in art, or religion. Our view on the whole shebang is tellingly told from the perspective of ordinary people, a new breed of smuggler, and we see events unfold from the point of view of these stalkers.

For the great mass of humanity, the Visitation passes by without a trace. In *Roadside Picnic*, man is described as a 'stationary system',[57] meaning that alien contact, as it is not a global catastrophe, has not changed the course of human history. Unlike *2001*, the elusive cosmic landing has not compelled man to take a different historical path, to be transformed into some post-human starchild. Indeed, the Strugatskys appear to be realists of the fantastic, portraying a potentially decisive moment in history with starkness, rather than sensation.

Even the alien motivation for contact is left in doubt. In voicing the view of the experts, Dr Pilman, the book's Nobel Prize winner as discoverer of the 'Pilman Radiant', suggests the root cause is the chasm between the civilisations, earthly and alien. And while this gap may be too great for terrestrials to surmount, the extraterrestrials seemed to have been reluctant to give Earthlings a shove up the cosmic ladder. Just a few fragments of their technological culture, a few mysterious artifacts strewn across the terrestrials' landscape can hardly be called helpful. Man is in the same position as mere woodland creatures rummaging among the roadside remains of a picnic.

Yet the artifacts are toxic, lethal to all living things. And rather than being thrown away in some little-used corner of the globe, they are pitched into the thick of it, into the densely populated district of a metropolis that represents less than one per cent of the Earth's surface. Though the universe has been tossing cosmic objects at the Earth since before the period of Late Terminal Bombardment, to date none has fallen on a city. And so perhaps this is not the work of chance, but of choice. We are left wondering if perhaps, rather than an indifference to humans, the aliens were more malevolent in their intentions.

Ultimately, perhaps there was no landing after all. An alien spaceship replete with a highly cultured cargo destined for distant shores stumbles into the Solar System. No one on Earth managed to see a single visitor because this is a drone ship, piloted by artificial intelligence and harbouring no biological life. In the vicinity of the Earth the ship sustains fatal accidental damage resulting in its breaking into six parts, which, like the debris of Comet Shoemaker–Levy, plunge into the planet from their orbit near Earth, creating the 'Pilman Radiant' effect.

The future alien

Maybe tomorrow, or maybe a decade or century from now, we may make the most shattering discovery of all time: the discovery of a thriving extraterrestrial civilisation. As the twenty-first century dawned, man had been imagining alien life for almost two and a half millennia. But as space agencies prepared to build flotillas of space telescopes to search for life in this unearthly universe, the crucial questions remain unanswered.

Today, the plenitude principle abounds. It was the influential philosopher and intellectual historian Arthur Lovejoy who first teased out the historical thread of a belief that the abundance of creation must be commensurate with an inexhaustible source. Lovejoy identified its philosophical origin with the Greeks, such as Anaximander, Epicurus and his follower Lucretius, a Roman. Though both idealists and materialists had used the idea of plenitude, it is Giordano Bruno who seems to have first inspired a Copernican take on the principle, his belief in an infinity of worlds leading to his untimely death.

The Church has had its doubts. At first sceptical of life beyond the Earth, it later produced theologians such as Thomas Dick whose zeal for extraterrestrial life led to an imagined Solar System population twenty-one trillion strong. And in the last century, it was writer Arthur C. Clarke who affirmed there was enough land in the sky to give every member of the human race, back to the first ape-man, his own private heaven or hell.

We live in a great age of discovery. We live at a time about which such writers and scholars could only dream. Astronomers, hunting for potentially life-bearing terrestrial planets around Sun-like stars, estimate there may be tens of billions in our Galaxy alone. A European team of scientists reported that perhaps forty per cent of the estimated one hundred and sixty billion red dwarfs in the Milky Way have a 'super-Earth' orbiting in a habitable zone that would allow water to flow freely on its surface.[58]

In a very real sense, the Copernican Revolution has been reborn.

The American space observatory *Kepler*, launched in 2009 to find Earth-like planets orbiting other stars, took off four hundred years after Galileo's first use of the telescope, and is of course named after the first great Copernican theorist, Johannes Kepler. Based on *Kepler*'s early findings, Seth Shostak, senior astronomer at the SETI institute, estimated, 'within a thousand light-years of Earth', there are 'at least 30,000 habitable planets'.[59] And based on the same findings, the *Kepler* team projected that there are 'at least 50 billion planets in the Milky Way', of which 'at least 500 million' are in the habitable zone.[60] NASA's Jet Propulsion Laboratory was of a similar opinion. JPL reported an expectation of two billion 'Earth analogues' in our Galaxy, and noted there are around '50 billion other galaxies' potentially bearing around one sextillion Earth analogue planets.[61]

Over the last two and a half thousand years, a stunning array of scholars, scientists, philosophers, film-makers, and writers have devoted their energies to imagining life beyond this Earth. Their task has been to try reducing the gap between the new worlds uncovered by science and exploration, and the fantastic strange worlds of the imagination. Their huge contribution has been important not only in the way that the fictional imagination has helped us visualise the unknown, but also for the way in which it has helped us define our place in a changing cosmos.

A mode of thinking has emerged. One in which the science and culture of astrobiology have been fused. In this rich evolution of the extraterrestrial life debate, science-fictional visions have influenced issues and dialogues in astrobiology, and in turn popular culture has been inspired by discovery and invention. The history of astrobiology has hinted at the revolutionary effects on human science, society, and culture that knowledge of another civilisation will bring. If we may be so bold as to suggest that humanity is at least a way in which the cosmos can know itself, what more is out there to be discovered?

Notes

1 Clarke, A. C. (1968) *2001: A Space Odyssey*, London, Preface

2 Lem, S. (1961) *Solaris*, trans. J. Kilmartin and S. Cox (2003) Croydon

3 Strugatsky, A. and Strugatsky, B. (1972) *Roadside Picnic*, London, Gollancz edition, SF Masterworks (2007)

4 Seifer, M. J. (1996) Martian Fever (1895–1896) in *Wizard: The Life and Times of Nikola Tesla: Biography of a Genius*, New Jersey, p. 157

5 Spencer, J. (1991) *The UFO Encyclopaedia*, New York

6 Corum, K. L. and Corum, J. F. (1996) *Nikola Tesla and the Electrical Signals of Planetary Origin*, OCLC 68193760, pp. 1–14

7 Janes, L. (2009) *Lem under the Lens: The Communication of the Science and Culture of First Contact within Stanislaw Lem's novels Eden (1959), Solaris (1961) and His Master's Voice (1968)*, Cardiff

8 ibid.

9 ibid.

10 http://en.wikipedia.org/wiki/Mary_Celeste

11 Quoted in Nunan, E.E., and Homer, D. (1981) Science, Science Fiction, and a Radical Science Education, *Science Fiction Studies*, **8**, 25, pp. 311–30

12 Lem, S. (1961) *Solaris*, trans. J. Kilmartin and S. Cox (2003) Croydon

13 ibid., p. 204

14 ibid., p.204

15 Quoted in Janes, L. (2009) *Lem under the Lens: The Communication of the Science and Culture of First Contact within Stanislaw Lem's novels Eden (1959), Solaris (1961) and His Master's Voice (1968)*, Cardiff, p. 43

16 ibid.

17 ibid.

18 ibid.

19 ibid.

20 ibid.

21 ibid.

22 Lem, S. (1968) *His Master's Voice*, trans. M. Kandel (1983) Evanston

23 ibid.

24 ibid.

25 ibid.

26 ibid.

27 Davies, J.M. (1990) *Stanislaw Lem*, Mercer Island

28 Swirski, P. (1997) *A Stanislaw Lem Reader*, Illinois, p. 42

29 Dick, S. J. (1996) *Biological Universe*, Cambridge

30 Youngblood, G. (1970) *Expanded Cinema*, New York, p. 139

31 Kaufmann, W. (1982) *The Portable Nietzsche*, New York, p. 124

32 ibid., p. 227

33 Rose, M. (1982) *Alien Encounters: Anatomy of Science Fiction*, Harvard, p. 32

34 ibid., p. 144

35 Clarke, A. C. (1984) *1984: A Spring of Futures*, New York, p. 111

36 Miller, M. C. (1994) *2001: A Cold Descent, Sight and Sound*, Vol. **4**, Number 1, p. 24

37 Burns, J. F. (1997) For Arthur C. Clarke, What is Paradise Without Praise? *New York Times*, 1 April 1997

38 LoBrotto, V. (1977) *Stanley Kubrick: A Biography*, London

39 Clarke, A. C. (1956) *The City and the Stars*, New York, pp. 174–5

40 Clarke, A. C. (1972) *Report on Planet Three & Other Speculations*, New York, p. 89

41 Clarke, A. C. (1990) *Childhood's End*, New York, p. 8

42 ibid., p. 8

43 Eisley, L. (1957) *The Immense Journey*, Random House

44 Aldiss, B. (1986) *Trillion Year Spree*, London, p. 401

45 Books of the Times: Two Tales for the Future, *The New York Times*, 22 August 1973

46 ibid.

47 Aldiss, B. (1986) *Trillion Year Spree*, London, p. 401

48 Lem, S. (1983) About the Strugatskys' *Roadside Picnic*, *Science Fiction Studies*, **10**, 31, pp. 317–32

49 Strugatsky, A. and Strugatsky, B. (1972) *Roadside Picnic*, London, Gollancz edition, SF Masterworks (2007)

50 ibid.

51 An empty consists of two copper discs, the size of Frisbees. No force on Earth is able to push them closer together, or pull them apart.

52 Sagan, C. (2000) *Carl Sagan's Cosmic Connection: An Extraterrestrial Perspective*, Cambridge, p. 183

53 ibid.

54 Stanley Kubrick Playboy Interview, *Playboy Magazine*, September 1968

55 Quoted in the movie *Contact*, directed by Robert Zemeckis, 1997

56 Wells, H. G. (1898) *The War of the Worlds*, Gollancz edition, SF Masterworks (2003) Chapter 1

57 Strugatsky, A. and Strugatsky, B. (1972) *Roadside Picnic*, London, Gollancz edition, SF Masterworks (2007) p. 100

58 The work of Xavier Bonfils of the Institute of Planetology and Astrophysics in Grenoble, reported in *The Guardian*, 28 March 2012

59 Shostak, S. (3 February 2011) A Bucketful of Worlds, *Huffington Post*, Retrieved 24 April 2011. Also at http://www.nasa.gov/centers/ames/news/releases/2011/11-07AR.html

60 Borenstein, S. (19 February 2011) Cosmic census finds crowd of planets in our galaxy, Associated Press. Retrieved 24 April 2011

61 Choi, C. Q. (21 March 2011) New Estimate for Alien Earths: 2 Billion in Our Galaxy Alone, Space.com. Retrieved 24 April 2011.

Index